BUILDING A GLOBAL BLOCKCHAIN BANK

BUILDING A GLOBAL BLOCKCHAIN BANK

How To Bank The Un-Banked Citizens Of The World

RABBIT HOLE CONSULTANTS

Rabbit Hole Consulting

CONTENTS

Give Thanks To Everyone!!!

History always proves that no journey is taken alone. And, there are people in my life who helped me in starting this journey as a author.

First of all, I want to thank the **Creator**, and all of my **Ancestors!!!** Especially, My Deceased Mother & Deceased Sister Mojuba Elegba & Egunguns!!!

Secondly, I want to thank my Aunt Dwaine for her eternal support, and loving commitment to family. You are a true asset in my life. To my brother, Thank You for the Mithril Order crypto-mixtape link. You gave me the soundtrack to the Blockchain Grind. Thank you Ms. Norma for making sure we had coffee in the final stretch of the book. Thank you to all of my family that have been there for me in the past.

I need to acknowledge a couple of friends who hung in there tight with the Blockchain movement. Hey Ifa Dare/Mr. Bailey, thank you for the 3 hour talk we had a few weeks before I finished this project. I needed that tune-up. We have to build something Blockchain together.

"Indeed, Indeed, Indeed...", to Metaverse Slim you already know...from the stockroom to the Blockroom!!! The Blockchain Bank gets set-up first, then the Independent Crypto-Record Label/Entertainment Company with the NFT's, Metaverse, Tokenized Performance Platforms, Digital Radio Station, etc...You know the real value that we can bring to the people, and culture. Thank you Sir!!!

Book Publisher

RABBIT HOLE
CONSULTING
BANKING THE UN-BANKED

Author Notes

Before we start this journey of "Building A Global Blockchain Bank", we need to discuss what type of journey we will be taking to achieve our goals. Also, we need to understand how this journey will unfold during our travels together.

The purpose of this book is to address a multitude of issues affecting people from different backgrounds around the world. Some of these people live in countries with highly developed financial infrastructures. Some of these citizens live in countries with rapidly developing financial infrastructures moving at blinding speed. But, the majority of the 8.5 - 9 billion people on the planet live in societies, conditions, and political arenas that prevent their access to stable financial markets. And, these people need access to the financial services, and products that will be essential for them to improve their quality of their lives. Their societies, and their future generations will also benefit from having access to financial markets. Everyone is not in the same place financially. What this means is multiple readers will be taking entirely different journeys throughout this book at the same time.

If you are a reader who lives in a country that is considered a "first class" economic society then you might be familiar with all of the aspects of the legacy economy. Or, you might be aware of the legacy economy, but you could be limited, or even prohibited from accessing the benefits of these financial services & products for various reasons. Those citizens that reside in rapidly developing countries with emerging economies will have a journey of their own making. These are the people who are currently incorporating, and experimenting with new financial services, and products on a daily basis to improve their lives. But, the majority of the world's population will have the longest yet most rewarding journey of all of the readers. These are the people who have never had any access, exposure, nor experience with financial products, and services from the legacy economy. So please be mindful as this book unfolds that this journey is a shared highway that will have various different entry points, and exits that readers will take as they see fit. Nonetheless, all travelers are welcomed to take this journey.

Before we get into the actual contents of the book let's discuss the format of this book, and what you can expect. Each chapter will begin with a brief description of the topic that will be discussed. This will make the book function as both an onboarding, and reference guide for any future use as a guiding point. A basic definition of each legacy banking sector, or aspect will be given after the chapter description. The same defining practice will apply to the Blockchain functions, and protocols as needed. Once we have a description of the chapter, and a definition of the financial aspect that we will be discussing in the chapter then a explanation of the current state of health for that financial utilitity will be given to the reader. This will be followed by explanation of how a switch over to Blockchain technology will provide benefits, and resolutions to the problems presented by legacy banking practices. Following the lists of Blockchain benefits will be an actual list of Blockchain Protocols for the readers to research, explore, and use to replace that particular legacy banking sector, or banking department. We will be closing each chapter with a brief summary of both the chapter, and the banking aspects thus far covered along with our bank building progress.

Legacy banking already has the infrastructure to build a bank. We will not be re-inventing the wheel, but we will be replacing the frame & all of the spokes. For every legacy banking aspect, and financial department that we use can be replaced with various Blockchain protocols to improve its functionality. The piecing together of these various Blockchain protocols will give us a functioning decentralized global bank built on Blockchain technology. Step by step we will systematically progress through the various aspects of the legacy banking to address the problems that they present to modern society. This will be our process of financially engineering a "Global Blockchain Bank"!!!

Preface

For the past few decades people in my life have been asking me when will I write...the book. Most of the time they were asking me when will I finally write that book about my crazy uncanny journey in life. I usually respond sarcastically to the question to deflect from the request, and change the topic. "I'm a Scorpio, I don't like people in my business...I was raised not to speak to strangers...people are stupid, they don't read anymore...". I have a whole stack of rebuttals ready, and waiting to be fired off at the next request. I haven't lived the greatest life in the world, but the life I have experienced is definitely esoteric, adventurous, and multi-dimensional to say the least. As I grew older it was easy for me to see that I'd never have a normal life like my friends, and other family members. I was born unto a mother, a woman, and a parent who forbid her child to lie to her, or anyone else for that matter. And, she did not lie to her children in return. I got beatings for telling lies not for what I did wrong. So, I grew up always needing to know the truth in all situations no matter how small, or grand the subject was to me. I tried my share of lies, but I didn't want to be fake like the other people. I wanted to be the smartest one in the room, the truth, and the most trusted. The foundation that my mother laid down became the corner-stone of my character throughout my life. Most people never pray for the truth, but I did so life revealed to me a totally different path that had to be traveled. It was explained to me as a child that the only difference between my life today, and my life 5 years from now is the books I read, the people I meet, and the places that I travel to in that time period. I needed permission to go out to meet new people, and to travel to other places. But, I didn't have to ask anyone to read a book. I found out that I could sidestep normal boundries, and experience a life that I wasn't allowed to see for myself. I read about card tricks, illusions, criminals,

Africa, slavery, spies, weapons, wars, sex, martial arts, science, animals, money, music, etc. The more I read the more I wanted to experience it in real life. Before I knew it I ended up with a life that nobody was able to guide me through because I was living a life most didn't even know existed. After reading thousands of books it was obvious that I was alone in my thoughts, and experiences. The sharing of these thoughts, and experiences in discussions is what usually sparks the conversations about me writing a book.

People in my circle know I have an alarming addiction for reading books that I got from my mother. If you hop in my car then you will probably see the backseat filled with books that catches your eye. When you pop the trunk you'll see another dozen books thrown in the back just out of arms reach. See I grew up in a household where if I said I was bored, or that I didn't have anything do I was told to read a book, a dictionary, a magazine, or the encyclopedia collections that filled our home. In fact, every time I asked how to spell a word I was told to go get one of the dictionaries from the book shelf. This was my life until I reached my pre-teens when tragedy struck our household, and my mother, and unborn sister was killed. Then I became a child that didn't care about much in life including my high school education. Although I was always one of the smartest in my classes, I refused to show it unless a competition was involved, or I had something personal to prove. Needless to say, coming from a family of educators that expected me to become the valedictorian this was extremely alarming to the rest of my family. But, losing my mother and sister left me at a crossroad where everything was in question including my future life, family, religion, and post-secondary education. I'm probably the last person that my family, and childhood friends would expect to never graduate from college. Not graduating from high school with honors, nor graduating from college left me with a subconscious feeling of shame, and a grandiose need to compensate. This need to compensate is what propelled me back onto a path of reading books that number well into the five figures. Even today it is almost impossible for me to read just one, or two books at a time. My childhood reaction to family tragedy, and my passion for deep research in every aspect of life is the reason that those who know me have been expecting me to write books for the past few decades.

Well this may not be that book about how adventurous my life has been thus far, but none the less it is the first book that they have been waiting for me to write. And, it is a book of adventure, and deep passionate research about both my life, and their life. This book is about my Blockchain life, and the future of their Blockchain lives. This book is a consolidation of almost 15 years of my research, experience, and travels throughout the Blockchain ecosystem. I've watched Blockchain technology evolve from a mere concept on a whitepaper, and clear-net website to becoming the backbone of the first successful global digital currency. I've seen Blockchain expand it's capabilities beyond a digital currency, and adapt to the needs of the people, the financial markets, and various current events around the globe. Not only have I seen Blockchain adapt, but I'm watching the entire world adapt to Blockchain. As of now, Blockchain is the cornerstone for a global financial paradigm shift that includes digital currencies, a new digital asset class. It's an open-sourced decentralized financial platform accessible to anyone with an internet connection. Plus, it's a freedom technology in the palm of your hands. If you have a mobile device with an internet connection then you meet the minimal requirements needed to access this open-sourced global financial ecosystem. This access via Blockchain will give you the same opportunities as a wall street investor has living in the heart of the New York financial district.

I have been researching, and studying the history of money & banking for decades. I was a child when I realized my aspiration to have my own bank long before I would realize how I would achieve this goal. Even as a child playing monopoly it was more important to me to be the banker than it was for me to be the owner of Park Place, and Boardwalk. After a dozen years of using Blockchain in various ways it occurred to me that all of the pieces were right in front of my face waiting on me to realize my lifetime goal. I had seen the silhouette of the potential, but not the full boldness of it's interior workings. I had spent to much time trying to convince non-believers of the benefits of Blockchain, and not enough time putting the pieces together. All of that changed one day when I was working for a Global 100 corporation named Wipro in 2017. I was working on multiple projects in their Business Processing Services division while conducting long debates about Blockchain to convince myself that a major change in my life was on the horizon. Ironically, it was one of my co-workers who had rolled their chair over to my desk to give me my flowers on a day that I was bogged down in my work, and being anti-social. They basically said, "Wow, you were right...I know you seen that email they just sent out to every employee...I know your going to be the first one to signup for the Blockchain certification once Wipro puts it together...". The one day that I was fed up with checking my emails Wipro had sent out a company email to over 100,000 employees asking if anyone would be interested in taking a new certification on Blockchain in the near future once they figured out what is Blockchain. The global giant Wipro didn't even know what should be in the Blockchain course, but they knew they had to prepare their workforce to provide these future services to Fortune 400 companies. I had been experimenting with Blockchain for about 6 - 7 years, and Wipro was just now teaming up with IBM, and B9 Labs to figure it out. When I took the certification course I had realized they really don't know a damn thing about Blockchain. It was at this point that my personal interest became a full-time professional crusade to become a Blockchain Professional. Later that year I was deployed as a Federal Emergency Management official performing disaster relief for Texas, Florida, Puerto Rico, and the U.S. Virgin Islands in response to Cat 5 hurricanes Irma, and Maria. While working for the federal government I had joined a couple technology list-server groups, and quickly realized that I was the only one who had Blockchain experience, and was working on becoming certified. Various governmental workers kept reaching out to me for information to explain what I knew about blockchain. Without knowing I had become a Blockchain consultant. I just wasn't getting paid for all my consultant work yet. After returning from the Virgin Islands I joined a new meetup group called the GBA (Government Blockchain Association) in Atlanta, GA. I can thank Randall, and Gerard for providing an additional space for me to assess my Blockchain acumen, and where I was in my journey compared to other people interested in this new technology. It was during these years that I started my Blockchain Consulting business (Rabbit Hole Consulting) out of Estonia using their E-Residency program. I had chose Estonia because it was the only country to actually embrace the Blockchain technology, and apply it nationwide.

Once I started Rabbit Hole Consulting out of Estonia, I felt even more compelled to know everything about Blockchain. So I decided to map out every ecosystem, all of the new emerging sectors being developed in the Blockchain realm. I had built a Blockchain course on the platform Teachable, but felt like I was disrespecting Blockchain by hosting it on a centralized platform. So, I started researching all of the Blockchain-based social media, and community platforms. I wanted a totally decentralized platform free of any centralized control. I was looking for something pure. It was during this process that I began to realize that my goal was attainable. This is how I realized that all of the pieces to create a Blockchain Bank were in place. I just had to wait for the various pieces to grow, and mature to the point where it was safe enough from a security standpoint. Then I could feel comfortable to put my name behind this new Blockchain banking entity. From 2023 - 2024, I started actively mapping out all of the Blockchain protocols that could actually replace legacy banking functions. Once I had my map then I thought it's time to write the Whitepaper to show my proof-of-concept. Because that's what everyone does right? Then you write your business plan, and figure it all out. Well, I'm not like everyone else, I don't follow others all like that. And, if your name isn't Andreas Antonopoulos then I haven't been following you at all in my Blockchain journey.

In 2024, I began consolidating several years of research, and experience in this book. I decided that this book would be the whitepaper, the on-boarding guide, and my business plan all-in-one. I decided to use the book as a way to both sum up the past, and map out the future.

In addition to, almost 15 years of studying blockchain this book also includes decades of deep research into how both the legacy financial system works, and doesn't work. In the past couple of decades while the Blockchain ecosystem was being birthed, I've realized that the legacy economy was rapidly dying off at an alarming rate. One of the most horrifying truths of the death of the legacy economy is the fact that the majority of the people on the planet have never even had access to it's resources, or opportunities. The global population is between 8.5 - 9 billion people, yet less than 20% of the earth's inhabitants actually have access to financial tools, plus the acumen to use them properly. This is more than just neglect, and backwards...this is by intentional design. This is Evil (live spelled backwards)!!! Financial technology is the foundation, and the conduit to improve our lives, our societies, and our modern world as whole. Most of the modern societies nowadays are structured on top of their financial systems, and their ability to wield these systems as they see fit. We have reached a point in our societies where financial power outweighs political power in every arena of life. This is my attempt to level the playing field. The technology to achieve this goal is currently available, and the pieces are right in front of our faces. Most people do not even know this opportunity is available to them, but they need to be informed before the window of opportunity closes. The goal is to bank the un-banked people of the world. The technology is called **Blockchain**. The opportunity is also called **Blockchain**.

In a way this book, ***How To Build A Blockchain Bank*** is a combination of a white paper, a proof-of-concept, and an on-boarding guide. All of the financial pieces are available to create our own banking entities with Blockchain technology. This book will function as a proof-of-concept by explaining the basic steps to manifest a Blockchain banking platform. The presentation of this information will enable people to create their own customizable banking services to improve their lives. Thw book also serves as an on-boarding guide with definitions, resources, and Blockchain protocol listings for different financial products. Whether you are interested in the concept of Blockchain banking, the actual building of a bank, or just some of the financial services rendered this book will serve your purpose.

This book is also a canary in the global financial mine shaft signaling the death of the legacy banking system. Consider it a signal, and a warning to guard your financial well-being. And, use it as another weapon in your arsenal to defend yourself, your families, and your communities. In the current climate of financial wars this is just another trumpet announcing the opening of another front on the economic battlefield. Throughout the history of warfare the final outcome has always come down to the introduction of a new technology into the arena of battle. In the spirit of *"Wargames"*, this is how I am burning down Joshua's tree. Blockchain is the backdoor that I have found that will help me escape Falken's Maze. The maze that I am referring to is the collapsing global fiat economy, and the calamities that will persist with it's coming collapse.

Introduction

The Death Of Legacy Banking

And The Coming Collapse Of The Legacy Economy

Making The Case

A case needs to be made for the death of the legacy economic system before a case can be presented with evidence for a new financial system based on a Blockchain economic model. This is a difficult task, and the goal of this book is to both explain the death of one system while explaining the birth of a totally new yet foreign system. The difficulty of this task stems from three factors. The extensive history of the legacy economy that we are currently employing. Plus, how do you explain the death of everything we know, and the birth of something that is unknown. And, where do you start making the case for the coming collapse of the global modern economy. To solve this problem I turned to a cultural ritual that is familiar to most people from various diverse backgrounds. I believe the best way to discuss the life, and death of anything is with an eulogy. It is with the use of the eulogy model that we will proceed to make the case of the death of the legacy banking system. It is in this fashion that we will be able to reveal both the history, and the demise of the legacy banking system. Every eulogy includes hope for the future through the offspring that were left behind. And, Blockchain is the motherless child that has become the new pretender to the throne.

We are currently living in the Age Of Purgatory. We're living at a time in which the old world is dying off, and the new world is still being birthed. There are a lot of lost souls roaming the planet aimlessly. The weakest ones feed off the energy of the masses to mask their capitulation to their own personal demons. This is true for every society, and every member of society regardless of their status, or self-perceived importance. Nowhere is this more obvious than in the world of economics. Modern day pharaohs create demonic economic policies to convince the masses to waste their entire lives to build timeless institutions in the pharaohs name. And, seeing that modern society is built on a pyramid structure it would probably be better to begin the eulogy of the pharaohs first, and foremost. Here are some recent events from modern society's economic pharaohs.

In the third week of May in the year 2024 a financial dog whistle was blown for 24 hours straight without a pause. It was blown so loud that even the youngest of the Gen Z'ers were barking out loud on the various social media platforms. And, my generation which is Generation X immediately recognized that sound as being the sound of the captains of finance screaming "Abandon The Ship Mates!". If you do not know who Jamie Dimon, Klaus Schwab, and Martin Gruenberg are then you probably don't know what they have in common. If you are familiar with these three names then you know they're the heads of some of the most powerful economic entities on the planet. Jamie Dimon is the CEO JPMorgan Chase. One the most powerful banks in the United States, and the world. Klaus Schwab is the founder, and executive chairman of the World Economic Forum. He has held that position for over fifty years. Meanwhile, Martin Gruenberg is the Federal Deposit Insurance Corporation Chair. It is the FDIC that insures U.S. banking deposits up to a certain limited amount. All three of these individuals are major leaders in the global legacy economy. But, that is not the commonality that I am highlighting. What I am bringing attention to is the fact that all three of these major captains of finance announced their retirement within the same 24 hours. The man who has run the most powerful banking entity in the United States. The man who is responsible for providing insurance for everyone who has money deposited in U.S. banks. And the man who had set up an economic forum for the world said it's time to quit on May 20th & May 21st of 2024. I'm going to leave the calculation of those odds to the Wall St. quants.

Just a few years earlier in the beginning of the third quarter of 2020 another economic milestone of recognition had occurred. In fact, October 1st, 2020 is a date that serves as a guide stone pointing out the direction that the health, and direction of the legacy economy has taken. This date is also a major reference point in my personal Blockchain journey. It is a date that can be used to diagnose how sick the legacy banking economy has become to date. It is on this date that Donald Drumpf the man who was occupying the seat of the presidency announced that he had contracted Covid-19. The announcement came late in the evening on a Thursday. By Friday morning when financial markets opened there was widespread panic in the legacy economies for both local, and global markets. The Dow Jones had plummeted around 400 points, the S & P had loss between 1.5% - 2% of it's value, Nasdaq losses were closer to 2.5%. On top of that the airline industry was in total panic mode because airline shares had loss on average around 3% each. The only reason the airline shares stop tumbling was because leaders in congress publicly declared that they were prepared to provide a financial band aid to stop the bleeding. These were the losses within "the strongest economy on the planet". But, globally countries, and their GDP's had loss between 3% - 10% of their values. Because it was a Friday that meant that two bells had rang that day on Wall St. The second ringing was the saved by the bell sound. Which came in the form of the weekend. Because the legacy economy shuts down the markets on the weekends that meant the panic was halted by default. Another interesting fact about October 1, 2020 is that most economist in the legacy economy use this date to push a false narrative about how robust, and resilient the legacy banking economy has become since its rebound. But, the undying truth is that 574 banks have failed between October 1, 2020 - June 1, 2024. With another 63 banks on the FDIC's 2024 list of banks in the process of failing.

Meanwhile, the panic in the legacy financial systems meant something totally different to me, and everyone who understood the Blockchain economy. "TO THE MOON, BABY...TO THE MOON!".

Does The Death Of Legacy Banking Mean The Death Of Bankers...Hmmm, Actually Yes It Does!!!

There is a phenomena that is occurring within the banking industry that mainstream media refuses to shed light upon. But, anyone who takes the time to study the last 20 years of the global banking industry will be confronted with some obvious questions. How come bankers keep dying at an unnatural, and an unprecedented rate of speed. Why are these banker's deaths concentrated in higher clusters of cycles during certain periods of time. What are the hidden meanings behind all of these banker deaths. And, what does this say about the overall stability, and integrity of the global financial system.

In the past two decades hundreds of bankers have been murdered, committed "suicide", or just disappeared. When you take a look at the stated causes of their demise, and their relation to financial scandals and the investigations associated with those scandals an ongoing pattern emerges. There is a long history of global financial scandals, but a few have both hasten, and cemented the demise of the legacy banking economy. The worst type of scandals are the ones at the highest levels of the global economy. These are the scandals that are committed by the same institutions that are suppose to "safeguard" the trust, and integrity of the entire economic system. The three most prolific scandals that I will mention that have the strongest relationship with global banker deaths are the Fort Knox/Treasury Department/Federal

Reserve being caught selling tungsten-filled gold bars scandal, the LIBOR price fixing scandal, and the FOREX scandal. A brief description of these three institutional failures along with a random listing of one hundred plus names of dead bankers with stated causes should be enough to make the case of the coming collapse of the legacy economy. Well, at least for those of us with common sense, and foresight.

For a long time it has been known, and/or speculated that all of the gold that is supposed to be deposited in Fort Knox has been stolen by the Federal Reserve private bankers with the collusion of corrupt public officials. This is not hard to believe because it's a public fact that the Federal Reserve has stolen the United States of America's constitutional right to coin it's own money! Now, they just print up worthless paper monopoly money at will, and sell it to the U.S. government for a fee. It is these accumulated fees that compose America's so-called National Debt. You hear the term National Debt all the time like it's not that serious. Politicians throw that term around like it's just another bill the other political party forgot to pay. Have you ever asked yourself these questions? When you owe a debt, or when someone owes you...usually you know who owes who, right? So, if America is historically the largest, and strongest economic entity on the planet until recently , then who do we owe? And, more importantly how is this National Debt being paid off? Ask yourself this, with all of America's economic successes why can't we just pay it off?

Well in the early to late 2000's the status of America's national gold supply was answered when a major financial scandal started coming to light. Not only did it involve individuals being defrauded, but multiple countries at the same time. To truly understand how global this fraud is, or how it facilitates the death of the legacy banking system you need to understand how countries settle debts with other countries. They don't accept fiat paper money for debts the same way as people. All countries require debts to be paid in Gold, you know Real Money! So in the late 2000's, when Ethiopia was paid by the U.S.A. in gold

Ethiopia deposited that amount of gold in their vaults until it was time to pay their bills to another country. When it became time for Ethiopia to pay South Africa they just transferred a portion of the gold that they had received from the Federal Reserve Bank in New York to settle their debt. Around the same time the United States was settling other debts with China. So they sent several tonnes of gold to China to balance the books. Well, this is when the game got REAL!

Someone in China decided to do their due diligence, and started drilling holes into the gold bars that they had received directly from the U.S. Treasury/Federal Reserve Bank of New York. China ended up discovering the same truth that Ethiopia had discovered when South Africa returned their shipment of gold back to Ethiopia as an unacceptable payment for debts. The truth that multiple countries were learning at the same time was that none of the gold was REAL! Tonnes, and tonnes of gold was being shipped around the world that was fake. The so-called gold bars were filled with tungsten, and sometimes steel rods. They were just gold-plated bars. In fact, a couple dozen people were arrested in Ethiopia because their government believed someone had tampered with their gold while it was in their vaults. Because, surely the U.S. Treasury, and the Federal Reserve Bank of New York would "never" shipped fake gold to anyone. Even though all of the gold bars had U.S. Treasury stamps along with Federal Reserve Bank serial numbers in-graved on the bars. Meanwhile, China didn't have any problems arriving at their conclusions due to the strict political system that was operating in their country. In china you cannot just say oops I'm sorry, and go on about your day like nothing has happen at all. China needed answers, and launch an intensive investigation into both the origin, and scope of the fraud. What they discovered was alarming to say the least. They had found evidence that the fake tungsten-filled gold scheme had started in the Clinton administration when Alan Greenspan was at the Federal Reserve. It was revealed that over one million 400 oz tungsten bars were commissioned from a refinery, and shipped to Fort Knox. It was soon realized that the global gold market was saturated with an excess of a half of trillion dollars in fake gold in circulation.

Not only was the U.S. treasury/Federal Reserve caught selling tungsten-filled gold bars, but the Bank of England was caught perpetrating the same fraud. The scandal was so alarming that the Rothschild family who made their family fortune off of gold for centuries went into damage control. By April 2004, NM Rothschild & Sons Ltd. had publicly announced that they were withdrawing from trading gold to review its operations. This let everyone know a major scandal was unfolding, they did not want their name associated with this particular fallout. Australia's Perth Mint was also a culprit that was identified in the fake gold scheme. Even Australia's ABC bullion dealer conducted an investigation, and published photo evidence online.

Not only did countries get defrauded, but a lot of individuals who dealt in the gold markets were discovering that they had been sold fake gold bars too. Some of these incidents even sparked FBI investigations among other investigations. It is the so-called investigation by the New York DA Morgenthau, a family member of the same Morgenthau's involved in the Black Eagle Trust/Federal Reserve Illegal Gold Bond Issuance Scandal after World War Two. The New York DA's office alleged investigation of Stuart Smith helped cover up the extent of the fraud by confiscating all of the NYMEX records relating to the fake gold transfers. Stuart Smith was the Senior Vice President of Operations at the New York Mercantile Exchange. And, as the senior VP of Ops for the NYMEX, Stuart Smith would be the person in charge of record keeping for every single gold bar that was traded on the exchange. This means he would have access to all of the serial numbers that were stamped on the fake gold bars along with their refinery of origin. Once the New York DA's office confiscated the records they mysteriously ceased the investigation. And Stuart Smith, well he mysteriously just disappeared, and DA Morgenthau saw no reason to investigate his disappearance at all.

So Stuart Smith is the first name on this long list of financial professional who have been murdered, committed "suicide", or just disappeared. The list is so long that it would require its own chapter in this book. I decided to list the names in a separate appendix towards the end of the book to mke it easier for everyone to conduct personal research of this phenomena. Actually, this list of names warrants its own book, and deep independent research. Maybe that will be my follow up book on the death of legacy banking. Some of these names are associated with the LIBOR scandal, some of them with the FOREX scandal, and some were associated with the Deutsche Bank's 2 billion dollar financing of the U.S. presidential candidate Donald Drumpf.

LIBOR is an acronym for London Inter-Bank Offered Rate. It was basically the average rate of interest that multiple leading banks in London would charge to lend financing to other banks. LIBOR was a creation of the City of London. Not to be confused with London, England. These are two different places. London is a city, and the capital of England. But, The City of London is a City-State with a two square mile area in the center of London, England. The City of London has it's own laws, and rules that they follow which is separate from the laws of England. Think of it as a tiny country within a country that operates in the same fashion as the Vatican City operates inside of Italy. The difference is the City of London is based on financial sovereignty instead of religious sovereignty. It operates as an offshore financial haven for big banking that makes it's own rules within it's city limits. This is where LIBOR was born. The use of LIBOR gained momentum in the 1970's as the interest rate for the offshore Eurodollar market. In plain terms LIBOR was the interest rate used to borrow U.S. dollars that were stashed outside of the United States. As more banks, and more financial services started operating in these markets the LIBOR rate became the industry regulated standard to create stability in the emerging global banking markets. The banks were "supposed to" submit the actual rates that they would pay, and charge to borrow or lend money to other banks. But, some of the banks had other ideas, and agendas.

The LIBOR scandal, and related investigations came about when it was discovered that banks were not submitting the correct rates to be included in the average calculation rate for the banking market. Around 2011 - 2012 it became obvious that banks were inflating, and deflating the rates that they were submitting for their own personal gains. Without the banks submitting actual rates there was no way to truly determine the actual health, and condition of the banking sectors. To put this in scope, when you have 50 - 60 countries, and the banks operating in their countries relying on this interest rate it becomes a global banking crisis. Especially when you have less than 20 banks colluding together to perpetrate a global fraud on the entire global banking system. There are multiple banking sectors, but you can look at just one to see how destructive this fraudulent scheme truly is, the Derivatives market. The LIBOR system was used to underwrite $300 - $350 Trillions in derivatives alone without consideration of any other financial products, or services.

You might recognize some of the banks involved in these scandals. They may be the bank that you rely on for your financial needs. These are the banks that were identified in the criminal LIBOR investigation:

1. JP Morgan Chase
2. Bank Of America
3. City Bank NA
4. Barclays Bank
5. HSBC
6. Lloyds Banking Group
7. Deutsche Bank
8. Credit Suisse

9. UBS AG
10. Royal Bank of Canada
11. Royal Bank of Scotland
12. Bank of Tokyo-Mitsubisi UFJ
13. Sumitomo Mitsui Banking Corporation Ltd.
14. Societe Generale
15. Rabobank
16. Norinchukin Bank
17. Credit Agricole CIB

These revelations about interest rate fixing was first discovered in 2012. And, in 2013 another global banking scandal became public knowledge, the Forex Scandal. FOREX stands for foreign exchange. The Forex market is where foreign currencies are exchanged into other currencies. And, the FOREX Scandal involved banks colluding to fix the foreign exchange rates in the currency markets. An average day of trading currencies in the FOREX market is around $5 trillion dollars a day. Just let that sink into your mind for a second before we move forward. $5,000,000,000,000 a day, and the banks were, or are manipulating the price of the exchange rates at the expense of everyone else operating in the FOREX market.

So, the gold of entire national treasuries is being stolen, counterfeited, and used to pay global debts. The interest rates for lending, and borrowing are being manipulated. And, the price of exchanging national currencies are being fixed by insiders. These are just three of the global banking scandals that is destroying the legacy banking system. The worst part is the fact that nobody on the planet actually knows what the actual value, and lack of value amounts to in the global derivatives markets. When the derivative markets collapses none of these scandals will come close to that calamity. Everyone within the financial markets knows that the bursting of the derivatives bubble will be a crossing of the Rubicon rivers.

It is time for the average citizens to stop acting like it's someone else's responsibility to manage your money, value, and resources. You cannot keep claiming ignorance. Nor, can you claim that you actually care about your financial futures until you begin to take control of your financial affairs. We need to build out our new world before the old world dies off. There were a lot of people standing on the deck of the Titanic watching the crew readying the lifeboats, but chose to keep partying because they didn't see other people preparing themselves for the coming disaster. It's time to jump ship, and build our own Arks before the next Great Deluge.

CHAPTER I

Legacy Banking As We Know It

What Is Legacy Banking? And what defines this banking system as legacy instead of modern?

The term ***Legacy Banking*** refers to what most people would refer to as modern banking, or our current global financial system. But in reality, it denotes an outdated decaying financial infrastructure that is no longer suitable for current, nor future financial facilitation. So when we refer to legacy banking we are describing a banking system that has been replaced by a new updated, and more current banking system infrastructure. This new financial system that has replace legacy banking will be unfolded, defined, and explored in the coming chapters of this book. When referring to this new financial banking system that is replacing the old legacy banking system, we will be using the term ***Blockchain Banking***.

So if the majority of people consider legacy banking to be our most modern, and current financial system then that raises a serious question. How can you call this financial system an outdated legacy? Easily, by understanding that the majority of people who use legacy banking have a lack of knowledge about how this banking system actually works. And, when someone doesn't know how the body is suppose to function then they have no way to tell when that body is sick, or dying until various systems start to shut down completely. When you do not have a working knowledge of your own system then you don't have a base-line to be able to gauge the benefits of other systems. Plus, a common trait of modern society is to cling to what is most familiar, and to reject what is considered foreign. Throughout history the most common fault of humanity is not recognizing transitions, transformations, and paradigm shifts. The current changes in our modern financial systems is equivalent to the fall of the Roman Empire. When global change was occurring it was the comfortable Romans who were the last to know the empire had fallen. They were going on with their daily lives totally oblivious to the fact that their system, and way of life had decayed to the point of total collapse. In a sense they were *zombies*. And, this is the current state of the legacy banking system filled with *zombie banks*, *zombie financial sectors*, and *zombie participants*.

Before we can diagnose the systems of legacy banking we will need to take a deeper dive into the various components that compose the "modern" banking system. So, let's start with the most familiar, and common aspects of our financial system. Then, we will work our way through the more complex sectors of the global banking system. Once we have identified the various aspects of banking then we will be able to see the pros and cons of each of these financial components. This will make it easier for us to see how legacy banking affects our lives. Then, we will be at a point to decide whether changes needs to take place, and in what forms that change needs to be manifested. Let's start at the origin of banking for the mankind.

The original fundamental reasoning for the creation of the banking concept was supposedly the dire need for an improved form of **Security** that an individual could not provide for themselves on their own accord. And, throughout various times in history this was a foundational truth that fueled the concept of a banking system. In fact, this fundamental truth of legacy banking can be traced back to one of our first documented legacy civilizations, Sumer. Before the advent of modern money when value was measured in perishable commodities resulting from each city-state's harvest, it was the Ziggurat temples that functioned as the secured banking vault. The Ziggurat temples were the most fortified structures in each city-state, and security of the temple was always held to the upmost regardless of harvest times. Only the high priests had the credentials, and temple seals to make deposits & withdrawals. An added security feature of the Sumerian Ziggurat temples were the facts, and beliefs that the temples were ruled by the Sumerian Gods. This would discourage even the most deviant members of the city-states from attempting a robbery of the temple-bank.

Thousands of years later this concept was popularized, and replicated by the Knights Templars. This replicated system by the Temple's Knights became the early foundations of the western banking system. Instead of using banking-temples the Knights Templars used their fortified castles that they had amassed in various territories as the banking facilities of the Christian empire. The culture of banking still has the feel of being inside of some sacred temple. People keep mundane conversation to a minimal, and accord each other a certain level of respect that is not exchanged outside of the church. And, if you are not there to worship then you are looked at as a potential sinner. Those who rob banks are seen as a moral threat to society. In our current times there are various reasons to look towards banks as a form of security. So security is the very first aspect of legacy banking that needs to be considered, reviewed, and assessed.

The next aspect of legacy banking is the most coveted yet the least understood. It is the most commonly used component of banking while being the most mysterious in both it's own nature, and human nature throughout history. If you ask the majority of people what is banking, or a financial system you will get an answer which is commonly known as the lowest common denominator, **Money**!!! But, if you ask the same people what is money, where does it come from, or where did it go, you will see the power that money has to manifest mystery instantly. And, this does not even include the question of what forms of money should we be using in our economy, and what are the best criteria that we need to consider when choosing a form of money. So, as a major cornerstone of the legacy banking system we will definitely be required to take the deepest of dives into it's mysterious waters. We will take a mature approach with a critical eye at the true relationship between modern legacy banking, and it's suspect relationship with the modern form of money. Once we complete our investigation into this relationship most of the readers will be ready to divorce legacy banking, and adopt a new form of money. This will be the point that most people will begin to seek out a marriage that truly takes into consideration their value, and the needs that they feel should be met.

One of the most common banking functions is **Cash Deposits**. To be more accurate the most frequently used banking function is depositing your accumulated value with others in the form of institutional deposits. In the agricultural age these deposits came in the form of perishable commodities. In Sumerian times the deposits were mostly grains. During the Mayan, and Aztec empires the cocoa was the embodiment of value. Now, in modern society there are various ways to deposit value in the legacy banking system. A person can deposit cash, or checks nowadays. People have additional options when you add on additional banking services such as safe deposit boxes. A safe deposit box gives a person additional options to deposit various forms of value. The use of a safe deposit box gives you the option to deposit jewelry, rare coins, gold & silver bullion, family wills, financial paperwork showing proof of ownership in businesses, personal property, and other investments. All of these sources are various forms of value that can be deposited in a bank. In a round about way they are considered as a form of cash due to their convertibility. But, in a more specific definition cash deposits refers to the currency on hand that you deposit via a bank teller, or an automated machine teller (ATM). In today's society less people actual make their deposits with an actual human. Most cash deposits, and withdrawals occur between the depositor, and ATM. Money can travel all around the world via banking wire transfers without touching a third-party's human hands, and give the depositor access to their money at an ATM.

It's not hard for most people to understand that they need to secure their money in some form, or fashion from a physical, or direct threat. But, when it comes to a more abstract view of securing your money a lot of us get confused, or even worst abused. What about securing your money from indirect threats. I'm referring to safeguarding it's value. Let's not confused money with value. They are two totally different financial aspects that sometimes work together, and other times conflict with each other's interest. What this means is the next cornerstone of legacy banking that needs to be assessed is **Store Of Value.** To be honest you can never have financial progress without a strong mechanism for store of value. Society loves to tell you whoever makes more has the most to spend, and a better lifestyle. Well you can have a lot of money coming into your various accounts, but if it is consistently losing value you don't have as much as it may appear.

As an example, let's take a look at two women who live in neighboring countries in North America. Both women have the same careers, the same number of bills, and the same amount of mouths to feed (5 family members each) at the end of their days. The first lady(A) is told she lives in a first rate country, has a first rate income, and lives a first rate lifestyle. The second lady(B) feels the same about her living situation, and lifestyle. Lady(A) makes $150/day & makes 3x the amount of Lady(B) who makes $50/day. Both women work a five day work week, and uses public transportation to save money. Lady(A) spends $5-$10/day to commute. But, it only costs Lady(B) $.50-$1.25/day to commute. It costs Lady(A) $10-$25 to prepare a full meal, but it only costs Lady(B) $2-$3 to prepare a full meal. Do you see where this is going, and understand who is spending the majority of their income. Hopefully, because we still haven't added in the cost of other bills (cell phone, utilities, clothing, entertainment, school supplies, etc.). Who do you think is able to retain a larger potion of their incomes. Who do you think actually owns their home property & vehicle. If this is a little too ethnic for you to grasp, then let's just keep it local. My grandfather bought his brick house literally for $3,500 with no mortgage, and no debt. His grandchild had to used that same $3,500 in the same currency as a rental deposit for a two-bedroom apartment with sheet rock walls, and neighbors on every side of him including above, and below his apartment. There was a substantial change in the value of the same currency that both my grandfather, and I used for our living needs. So, yes we definitely need to take a look at legacy banking, and it's relationship with our Store of Value.

For people the most common banking function is depositing real accumulated value into the banking system. But, for banks the exact opposite is true. Banks value a person's **Debt** above their deposited value. In legacy banking the banks profit off of the accumulated debt of the people. All real value deposits have financial limits. But, if a bank takes your deposits, and lends it to someone else then they can charge the other person an ongoing interest rate on your money that they lend to others. And, if the other person is not able to pay it back then the bank gets to collect their real value in the form of collateral, and add it to their profits. Debt is an agreement, or obligation to repay a lender who provided something of value to a person. In legacy banking these financial obligations are provided in a way that guarantees that the borrower will always pay a lot more than they asked to received. Let's think about this situation for a minute. A person, organization, or business has a need for more money than they actually have on hand. Then they ask for financial help from a bank to meet their needs. And, the bank's response is to make sure the borrower pays the accumulated interest rate first before they can start paying off the actual borrowed amount that they did not have in the first place. So now the borrower has additional financial burdens that they did not have before they sought financial assistance from the bank. When a bank creates a contract of debt the bank considers that debt as an asset of value. They actually record this debt in their accounting books as real value that they created. In case you missed what just happen I'll explain it another way. The more debt a bank creates the more profit the bank will make for the bank. The more profitable banks become the more indebted the people will become over time. And throughout history debt has always been a weak excuse for justifying slavery both financially, and physically.

In the summer of 2024 I found myself standing in the line of a credit union. On the wall beside the teller was a listing of services along with the percentages earned, and charged for deposits, and loans. It was my first time in this credit union so I started looking around like does anyone besides myself see this warning on the wall. I couldn't believe my eyes, and started saying to myself "this is why I am writing this book". The sign was color-coded with green highlighting deposits, and the interest that you would earn on your CD's (Certified Deposits). The bottom of the sign had two line items in red representing a personal, and a car loan. The green sections listed how many months, and years your deposit will need to be left in the institution to earn a certain interest rate on your money. The red section showed how much interest that you will have to pay on top of the amount borrowed, and the loan type. It did not matter whether you made a deposit for 6 months - 60 months because the credit union was only gonna pay you 0.1% - 4..5 % on your CD. But, when you looked at the red section it was more than obvious that you were going to pay at least 18% for both a personal, and car loan. Why are they charging different amounts of interests for the same amount of money borrowed. A loan of $10,000 is $10,000 whether it's for a car, or for personal use. Do the math...if I give the credit union a $1000 as a deposit for a year then the most I can earn is maybe $45= $1045, but if I borrow that same $1000 then I will have to pay at least $180 + $1000= $1180. And, this process keeps repeating itself as long as the deposits sit in the financial institution.

This banking debt scheme has become so profitable that banks have turned it into an exclusive club. The banks refuse to waste their time unless they know they can confiscate a significant amount of real value from a person. So to improve their profit margin, and time efficiency the banks have created a rating system to identify the most profitable victims. This is the basis of the **Credit** rating system. This credit system assesses a person, or organization's ability to repay a specific amount within a specific time frame. Over time this rating system has been fine tuned to weed out, or penalize those who can't repay in their time frame, or those who tend to repay too swiftly. Because when a person repays the bank at a responsible fast pace the banks feel that this person is not profitable. Legacy banks prey on the middle ground, or more specifically the middle class citizens. These are the people that have enough to repay, but not enough to repay all at once. You can look at credit as a contractual agreement based on a financial act of faith mixed with a financial assessment. Along with depositing cash, credit has become the backbone of a rotating system of give, take, and keep.

This revolving system of taking from one person (the depositor), and giving it to another person (the borrower) only to take more from the borrower (in the form of interest payments) is what legacy banking calls **Borrowing & Lending**. To the average working person to borrow means to receive something with the full intent of returning the item to the lender. For most people when they lend someone a $100 for a certain period of time they expect to be repaid that same $100. Not $99, and not $101 in repayment. But, for legacy banks it is a totally different definition, and situation. First, the legacy banks borrow the depositor's cash deposits usually for free. In this case borrow means the banks re-appropriate the value of others for their own use as they see fit to use that value. So that hard earned value that you accumulated, and placed with the bank for safekeeping was placed at the disposal of another person, or a business of the bank's choosing. Another debt-based contract was created, and secured with real world assets to secure that debt-contract. It was created with a time frame for repayment, interest fees, and other financial conditions attached to this debt-based loan contract. So the depositor's initial deposits that were given to the legacy bank to be secured was re-secured repeatedly by additional real world assets each time a bank created a new debt-based contract in their accounting books. And, as long as the depositors do not all withdrawal their deposits the banks can repeat this same scheme on an ongoing basis without any resistance, nor interruption. This is how legacy banking insures their profits.

When you have enormous amounts of excess capital/cash on hand then you are able to provide additional financial services to others. Another financial service, and product that is essential to the growth of economic systems is **Insurance**. There are several different types of insurance but, they all have the same basic functions at the end of the day. We've discussed how providing credit is an instrument of financial-faith. Well, insurance operates on the opposite end of the spectrum. Insurance can be considered a financial act of preparedness, prevention, and probability. Insurance is a financial product, or agreement that becomes active in the event of adverse events occurring. It's a contractual agreement to provide assistance, or reimbursement due to future loss of some kind. Look at insurance as a financial bet of probability that sooner, or later something bad can eventually happen to your business, property, or even yourself. In all aspects of life there are cycles of good, and bad that manifest in one's life. Insurance is a way to financially hedge against these negative manifestations. It's a financial means of covering certain liabilities that may, or will occur in the future. Insurance is a tool to hedge against these probabilities. In certain sectors of the economy having insurance is considered mandatory.

These banking sectors are some of the essential components defining the legacy banking system. There are various financial services, and investment markets that banking helps to facilitate. Banks have become business consultants, business partners, conduits cf currency exchange amongst other services. And, although they have worked wonders in the past their current statuses are now in question. It's not a question of if the legacy banking system is going to collapse. but a small question of when the collapse will happen. Another question is how many people, businesses, and societies will be destroyed by the collapsing of the current banking system. Now is the time to hedge against the coming liabilities. In other words, it's time to get some insurance. Blockchain is the insurance that I am proposing.

CHAPTER II

Benefits Of Blockchain In The Banking Sectors

Definition Of Blockchain Technology & The Term Crypto

One of the biggest obstacles to mass-adoption of the Blockchain technology is a clear definition of Blockchain, and it's common name Crypto. Some people think these two terms mean the same thing, and some people feel they have two totally different definitions. In a sense both groups are right, but a closer look at their relationship will provide a lot of clarity. At a basic level Blockchain is the backbone technology that is secured by Cryptography.

Blockchain is the public ledger of data, and cryptography is the means to secure all of that data. In it's simplified form, Blockchain is a public open-sourced digital ledger. This means there are no restrictions, and access is open to everyone. Blockchain keeps track of information, transactions, and proof of ownership. The basic steps that occur during the Blockchain process are very simple. Transactions made on the Blockchain are time-stamped with meta-data detailing the transaction's information. Multiple transactions are grouped together in bundles that are called blocks for easier processing, and bookkeeping. Each of these blocks are time-stamped, labeled, and encrypted with cryptography.

Each block will be attached to the previous block of transactions. This process is repeated all the way back to the original genesis block. This linking process creates a chain of blocks that holds all of the transactions, and their related information. It is this chaining of blocks that creates the **Blockchain**.

Because the Blockchain is a public-ledger, and open to anyone who needs to use it's technology, you will have various levels of sensitive information in the public domain. This means we need a way to secure all of the data that will be added on a constant basis. **Cryptography** is a means of secretly communicating information between two groups securely. One of the main aspects of cryptography is it's use of the hash function. The hash function guarantees that the same exact information/data that was encrypted will be the same exact information/data that is decrypted. Cryptography insures that the data can both remain public, and secure. This is why the terms Blockchain & Crypto are sometimes used to describe the same process, or product. The term Crypto is also used to refer to crypto-currencies, and crypto-assets that are built on top of the Blockchain technology. When Crypto is used in this fashion, you can think of it in the same way a smart phone, and apps function. As a metaphor, Blockchain would be the device, and Crypto would be the apps/applications added to the device.

Benefits of Blockchain Technology In Banking

Let's take a look at some of the problems that Blockchain solves so we can see both the benefits, and opportunities of Blockchain in general, and specifically in banking. The common denominator of legacy banking scandals, corruption, and failures is the centralization of power in the hands of a few private stakeholders. When the rules are dictated by a small exclusive group instead of the majority then major imbalances will occur without proper consideration for the majority's well being. Centralized power only benefits those at the top of the pyramid. Which means the rest of the foundation incurs all of the liabilities in a centralized system. It is this centralized structure of the legacy banking system that creates all of these multiple points of failure. As time progresses these failure points multiply, and only get weaker everyday. This creates a cycle of retardation because when concerns, issues, and problems arise there are only a few people with the authority to address the problems. Plus, the people at the top of the centralized structure are usually out of touch with the majority of the people in that system. This creates limitations in know-how, and the ability to act in a impactful manner. In contrast to the failing structure of the centralized legacy banking system, Blockchain is designed to be a decentralized open-sourced structure.

Blockchain's most dominant characteristic is it's decentralized designed. It is this decentralized nature that provides benefits, and capabilities beyond the capacity of the failing outdated legacy system. There are a handful of criteria that we as humans require to be free, and have healthy growth in our lives. It is the same criteria that is essential for positive financial growth, and well being. And, Blockchain provides both the capabilities, and the support in each of these areas. The five areas of consideration that Blockchain provides huge opportunities, and benefits are: **security**, **privacy**, **control**, **censorship**, and **scalability of development**. Let's take a deeper look into each of these aspects as it relates to financial facilitation.When we compare Blockchain Banking's new structure with the old outdated legacy banking structure the benefits, and problems will become evident.

First of all the financial system needs to be secured on an ongoing basis. So **security** needs to be able to keep up with both current demands, and future growth of technologies. Blockchain financial systems are built upon decentralized networks. A decentralized network does not have a single point of failure because both the managed-risk, and the data is secured by spreading out both the responsibility and information throughout the network among all of the participants in the network. When a centralized network gets hacked all data, processes, and networks are compromised at once. One point of failure can damage an entire centralized network. In Blockchain banking systems there are collective monitoring, governance, auditing measures in place for all participants. Any changes in a Blockchain-based system are voted upon through a public governance mechanism. Not in a smokey backroom by a few of central bankers for the benefit of a few investors. A decentralized Blockchain Banking system is more secure for both the investors, and for the network as it grows.

On the opposite end of the spectrum in-security in the legacy banking system is an accepted norm, and an industry practice at this point. The failed state of security in legacy banking is so prevalent that society as a whole believes that financial insecurity is a cultural normalcy. Nowadays, everyone expects their accounts to get hacked. Even banks, and businesses write these losses into their annual budgets. Because legacy banking is centralized that means virtually every aspect of their platforms are owned by a different business. Each of these businesses have a different degree of security posture. There is a continuously expanding landscape of attack vectors via the online, and mobile banking platforms. Websites are vulnerable by design, and third party plugins. Mobile apps are more concerned with collecting, and stealing customer data on behalf of the banks. They rarely make it their focus to secure the transacting process. Most mobile apps that are secure will still navigate the user to insecure links throughout the platform. Legacy banking technologies are usually facing both budget-ary, and time constraints during the development process. This is the culture of the legacy banking system.

The **privacy** capabilities of Blockchain banking systems is unmatched by the legacy banking system's predatory policies of user data collection. Blockchain banking offers anonymity during transac-tions while still providing a permanent record of the transactions for accountability. This same technology provides the option to link these same transactions to your real identity when there is a need for a personalized set of records. Blockchain banking gives you the opportunity of owning your own data, and the freedom from having your living habits mapped out by legacy financial institutions. In legacy financial systems banks function as data brokers that track your purchasing habits, and then sell that data to any data broker, or business entity willing to pay for your data. Blockchain banking gives you the ability to have a pure financial experience without third-party interference from people who have nothing to do with your transactions.

Living in the information age is probably the most intrusive time in human history. Never has there been a time in which a person's living habits has been mapped out by private entities. Our data is being collected at an alarming rate. And, the biggest culprits of data warehousing is the modern financial industry. Big banks collect big data. The banking industry collects more data than banking fees. They are the biggest data brokers on the planet. Not only do they track when you visit their platforms, but they track where you are geographically when you visit their platform. Often they will block a customer's access to their own money because they tried to login from a different location than the location data that they previously collected. Legacy banking routinely informs customers that they are not logging into their accounts from the device that the banks have on record. Who the hell are they to tell a private citizen which one of their devices the banks feel they should be using to access their own money. Think about how much that violates a citizen's privacy. Banking institutions has made violations of privacy an institutional culture. You do not have this level of privacy violations when transacting with Blockchain technology.

In the information age, and the current geopolitical climate it's important to understand the power of personal **control**. Not only is it your right, but it is also your responsibility to control your financial decisions, actions, and access. Blockchain banking gives you the ability to control the entire process of transferring value, data, and assets without third party interference. Through Blockchain banking these transactions occur without any third party involvement in the complete process. In a legacy banking system it is a common practice for third-party companies to restrict access to financial services, products, or even your own money and assets. The issue of financial control does not stop with just personal transactions when using Blockchain banking. A decentralized financial network prevents anyone, or any entity from controlling the banking network. This guarantees open access to all.

On a global scale people all over the world are losing access to their own accumulated wealth regardless of its amount. The age of financial custodianship is dead. It has become a common banking industry pratice to restrict access, and confiscate the deposits of institutional customers. Legacy banks are constantly depriving people of their legal, and financial rights in every manner that they can conceive. It starts before you even become a banking customer. The problem begins by attacking your rights in the ofuscated terms & Agreements contract. Predatory banks deceive customers into giving up all of their personal rights by having you agree to several different legal bindings via a check a box. With one click of the box...you agree to be violated at will by your new account providers. Most of the time you have agreed not to hold the bank liable for any future wrong doing on their part. All over the world mobs of people in various countries are raiding their local banking institutions attempting to retreive their stated deposit amounts. In some countries it's not safe for bankers to walk the streets, or be seen in public. Don't assume your bank will not try to assert full control over your personal wealth for their own financial benefits.

Blockchain banking offers a decentralized public-governance financial model. This is a **censorship** resistant financial system based on collective decisions. The Blockchain is an append only protocol meaning that only new data blocks can be added to the current chain of data. Blockchain does not have a re-write, or erase function within it's protocol. You cannot alter the on-chain data to benefit yourself financially, legally, or politically. No one can change the data against others because of their stance on race, politics, or ideology. Unlike legacy banking, there is no central authority that can censor, ban, or delete your content. And, this censorship free content includes all of your financial decisions, and actions that you perform in the Blockchain banking system. Now, there is no need to worry about being targeted financially due to factors that have nothing to do with your financial business.

Blockchain protects you against false accusations, character attacks, and contrary political motivations. Blockchain technology prevents politically motivated financial attacks on your well being via economic sanctions, and/or penalties. You cannot punish someone financially when you do not have total control over their financial operations. The world will become more peaceful when people stop forfeiting the sovereignty of their economic responsibilities. Once you give up control then you give up a decision process that benefits the needs of you, and your family.

All centralized legacy built structures suffer from the same problems. And, a lack of **development & scalability** is on the top of that list of problems. Centralized networks including legacy banking networks tend to have short life-spans because they lack these two essential criteria. In centralized systems decisions are made by a limited few individuals who are constantly struggling to grasp the ever changing needs of that system. Information has to travel vertically through several levels of gatekeepers before it can reach a decision maker. By the time the information reaches a decision maker it needs to be updated to reflect both the changes, and lack of changes in the system. In addition to the lack of adaptability, the decisions regarding the centralized networks are determined by the limitations, and acumen of a few restricted administrators. These same limitations restrict centralized networks from scaling their growth in truly meaningful ways.

But, Blockchain banking is an open-sourced network that encourages everyone to get involved in both the management, and development processes. With Blockchain banking being build upon a decentralized network this enables a multi-lateral distribution of information, and opportunities. The more contributors invest their collective time, acumen, and resources the more agile, scalable, and secure the decentralized banking system becomes with each contribution. Having a group of minds jointly working to identify, and solve problems can manifest new opportunities that an individual leader cannot conceive. A decentralized network will always have the human, and material resources that are essential for organic growth on a massive scale.

On top of all of these benefits one of the best benefits of Blockchain banking is that it is easily accessible. As long as you have at minimum a mobile device, and an internet connection then you will have access to the Blockchain banking ecosystem. So once people learn all of the various types of financial transactions that they can perform within the Blockchain financial system then a true paradigm shift can take hold of the global economy. Now, it's time to start thinking about how we can actually make this transition from legacy banking to a more beneficial Blockchain Banking system. It's going to take ongoing financial literacy to strengthen our financial acumen. And a commitment to become our own financial stewards. Lastly, we will need a clear understanding of what Blockchain pieces are needed to replace the outdated legacy banking functions, and sectors. In the next chapter we will begin learning how this process will take place in a common friction-less manner.

Here's a quick look at some innovative ways that Blockchain technology is being used to build products, services, and businesses with the utilities of NFT's:

Digital Art

You can create your own digital art. You can mint a NFT of a physical piece of artwork.

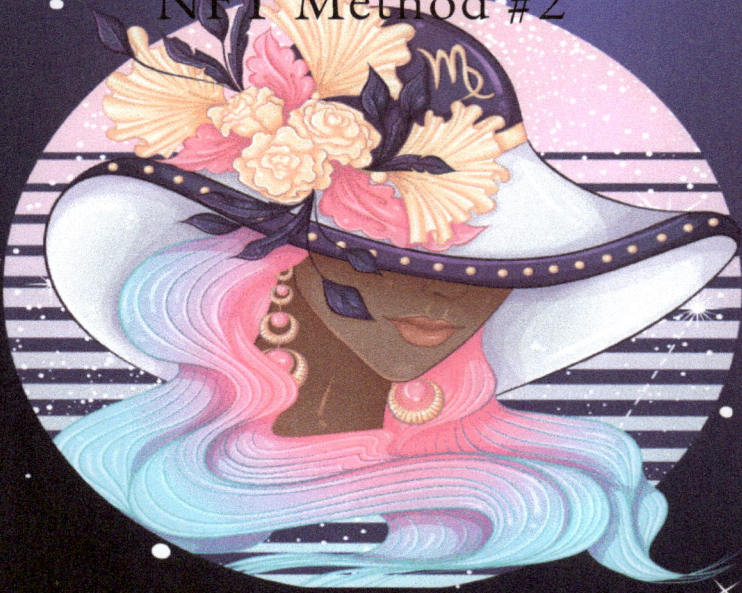

Royalties

Any NFT that someone mints can be created with up to 10% in royalty earnings on all future re-sales of that NFT.

This is a means of creating life-long sources of passive income for each NFT that you create.

Now Great Artwork, Products, & Services Can Generate Great Residual Income!!!

Fundraising

Organizations, Social Causes, and Campaigns can raise capital by distributing information, updates, and calls-to-actions in NFT form.

Funds for the Cause are generated as the information gets transferred throughout the organization's social network in the form of residual NFT royalties.

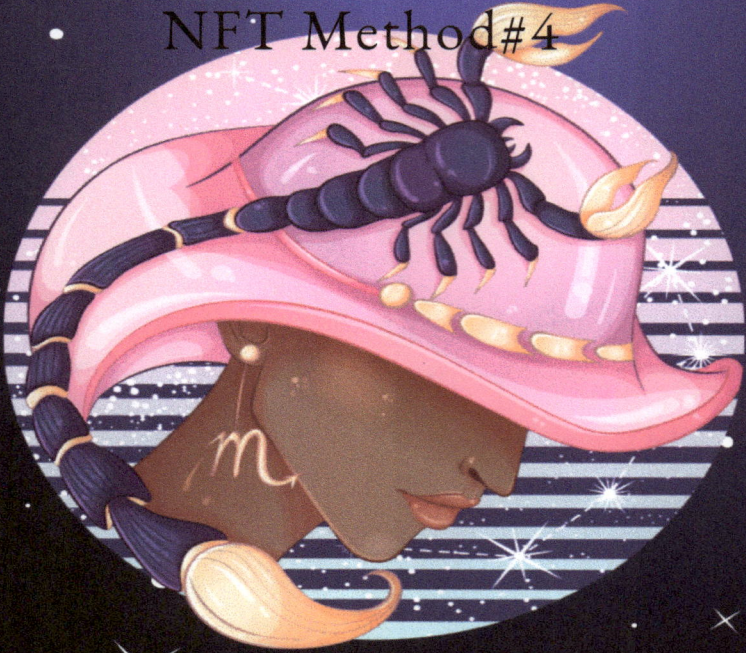

Music

NFT's are Re-Structuring every aspect of the music industry including these few elements:

NFT's provide a direct-to-the-market distribution model for both artists, and their supporters.

NFT's eliminate all artist ownership issues for artists with their Masters, Copyrights, Publishing, etc.

Blockchain-empowered NFT's enable permanent product tracking, and recordkeeping capabilities for sales.

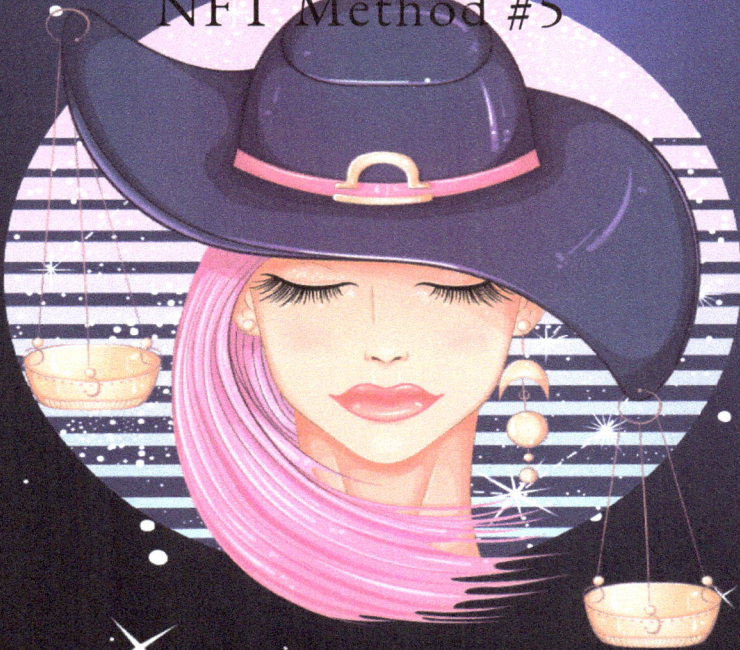

Collectibles

NFT's take collecting rare items to the next level.

Just as the collectible item's value increases, so does the NFT's value. Plus, the NFT is priced in crypto, so the price of that crypto-currency can increase in value.

NFT's can be used to authenicate, and document the rarity & characteristics of collectible items Such as:

Comic Books, Sports Cards, Coins, Stamps

Coupons

NFT's can be used to distribute discounts on real-world products & services. These coupons can be shared with additional customers. Giving businesses tracking data.

NFT's can be used to package product licenses, and establish affiliation programs.

Limited sales campaigns ,and limited product editions are perfect for the rarity nature of NFT's.

NFT Method #7

Content Creation

Use NFT's to start monetizing your content, and to create passive streams of income.

Use a Blockchain-based Social Media platform such as Steemit. This platform pays content creators in crypto-currency to curate valuable content on it's platform.

Blockchain social media platforms allow users to vote on the value of postings on the platform. Which directly determines the type of payout to the content creators.

Event Tickets

Use NFT's to design, distribute, and track event tickets.

Created Limited VIP Tickets

Verify Guest Lists & Identities with NFT's

Craft the ticket as an collector's item in theme with special events.

Use NFT ticket creation to manage ticket-market inventory.

Gaming

NFT's are used in Blockchain games as in-game digital assets.

These game assets can be used to enhance the gaming experience.

NFT gaming assets can be rented out, traded, and bought & sold.

NFT's are an essential aspect of the Play-To-Earn GameFi economic model.

Real Estate

NFT's can be used to invest in both physical, and virtual real estate opportunities.

With NFT's real-world real estate investments can be tokenized, and divided into fractional units of investment.

NFT's are also used to buy & sale virtual real estate in both games, and the Metaverse. Virtual real estate can be developed, and monetized just like physical real estate.

CHAPTER III

Replacing Banking Departments With Blockchain Protocols

Building Financial Departments With Blockchain Protocols

To the untrained eye a department in a financial institution can appear to be a complex process of indiscernible chaotic moving pieces. But, in truth a business department is just a grouping of people performing a set of assigned tasks, or processes in a designated space. Almost every business department can be broken down into 3 - 5 processes that help to achieve the company's business goals. All business departments start with one, or two people who are responsible for accomplishing a certain set of duties. As these people plan, and execute their duties their scope of responsibilities began to grow beyond their personal capacity. As the company grows it will need additional people to help manage the additional responsibilities that arise on a daily basis. The organization, and stacking of business goals, responsibilities, and personnel is the process of building out departments.

An example of a simple 10 - 20 persons department may be as straightforward as a pair of co-workers being responsible for two to four business processes in the entire department. A department may be responsible for managing only five business processes for the company. When two people are assigned to each of these processes you'll end up with a business department that has ten personnel members working together. So when you start planning your financial departments always start with the business goals first then determine how many business processes are needed to meet those goals. The number of people in each department is the last thing to consider. For each department list the business goals that you are grouping together. Determine what procedures need to be in place for the business goals to be achieved. Once you know how much work needs to get done then you are ready to decide who is the best persons to manage these departmental responsibilities. Then you can add personnel as needed. Lastly, the department needs a set of clear communication rules to facilitate information throughout the department.

In our modern connected world we establish communication between both humans and machines, and machines with machines through established protocols. Protocols are the rules, and the standards for a designated system. A protocol is a set of procedures that governs the communication functions of a network. It is through protocols that we establish formats, and structures that determine the rules of how data sets will be manage within certain networks. These protocols lay out how the processing logic will operate when certain interactions occur. All systems of interaction operate according to various protocols. Some of the most common protocols in our online world are https://, URL's, and IP addresses. All of these are protocols that govern the way we communicate as a humans with the internet. The same holds true for Blockchain networks. In the world of Blockchain it's the use of protocols that determines the nature, and purpose of the Blockchain.

Blockchain is built upon decentralized networks so there is a heavy reliance on protocols that provides details about how to interact with the network. These protocols lay out the governance, and decision making processes that establish the benefits of network participation. Blockchain protocols are predefined regulations that provide requirements for processes such as building a consensus within a decentralized network. The Blockchain as a whole has it's protocols, and then individual sub-blockchains each have their own set of protocols that govern that particular Blockchain. Each of these Blockchains have a set of protocols that establishes it's functions, and determines how data will be transferred. Some common Blockchain protocols are smart contracts, network & node participation, and the reward process. Other common functions include staking, data & value transfer, and automation.

To start building our Blockchain banking access platform we need to first set the cornerstone with a Blockchain Banking business plan. We need to determine the business goals of our bank, and where we want to place our focus, and resources. Once we've identified the scope, and the range of the Blockchain banking functions then we can begin the process of replacement. This begins with decisions regarding the type of banking services, and products that you want to interact with through Blockchain. Understanding basic banking functions are essential to deciding the Blockchain protocols that you will need to assemble. This will answer the question of how many different Blockchains you will need to interact with to achieve your financial goals. The type, and amount of tokens that you will need to acquire to facilitate your financial operations will depend on the Blockchain protocols are needed for each service, or product.

The replacement process begins with researching the various Blockchain protocols that function in each financial sector. For instance, if you plan on conducting lending activities then you will need to research Blockchain lending, and borrowing protocols. Determine the

pros and cons of replacement by making comparisons between the
legacy banking functions, and the Blockchain banking protocols. This
process involves systematically identifying, and aligning the Blockchain
protocols with the desired financial processes. Then, you'll need to
make a list of all of the digital tokens the will be needed to operate your
Blockchain financial entity. You will use this list as a starting point for
building your Blockchain financial investment portfolio. Acquiring
these tokens first gives you a diversified portfolio for stability plus the
ability to generate additional forms of value, and digital assets by
generating financial transactions. The number of different Blockchain
protocols needed to build a financial department will depend on your
desired scope of operations. You will need to check which protocols are
automated, and which protocols will need additional management
requirements. An understanding of the scope of responsibilities, and
time constraints will help determine the order, velocity, and cycle of
your operations. You may need to launch your own token in relation to
your Blockchain financial operations. You must decide if your Blockchain financial platform will have it's own governance, and liquidity
tokens as part of your design. Before you launch your token you will
have to undertake careful consideration of the tokenomics, and it's
overall functions.

Researching Blockchain protocols is basically just like researching
Blockchain companies to potentially invest money in via token
acquisitions. It's the same process that holds true when researching
potential business investments. Make sure you look at the founders,
and their backgrounds. Take a good look at the technology so you can
understand the benefits, and cons of using their protocols. You will
need to understand how the tokenomics of that protocol will affect the
growth, and the store of value when operating within that particular
blockchain network. Make sure each Blockchain protocol has had their
security, and programming code audited properly by the Blockchain
community. Take note of any similar protocols that have additional
benefits, cheaper costs, or ease of use properties. Look at the robustness

of their network, and it's supporting community. The support of the their community can determine the value of investing in that particular Blockchain protocol. It is a simple process of replacing banking departments with Blockchain protocols. But the process is intricate, and needs to be repeated until the desired format has been achieved according to your financial designs. Once you have your financial Blockchain structure in place then you can begin to customize your design to meet your own preferences, and operational needs. As stated earlier, this process begins with your Blockchain bank's business plan. When the business plan is detail-orientated, and concise then the design, and Blockchain protocol replacement process becomes more simplified with each decision. This provides a smoother transition from legacy banking services to Blockchain-based financial services.

CHAPTER IV

Becoming The Bank- The Business Plan

Putting The Pieces Together To Build The Blockchain Bank

Building a Blockchain bank may sound complicated, and the technology may be new, but the process to start a successful business is still the same. We need to start with a basic business plan. A business plan for starting a bank based on our needs, and our concept. This business plan will need to be international in scope because Blockchain technology is not limited by geographic borders. Also, our goal is to provide financial access to those people who do not have current access to the closed legacy markets. This means our customer base will be from various emerging markets around the globe. Remember as long as an individual has a mobile device, and an internet connection they will be able to access Rabbit Hole Global Bank's financial services.

Let's start with the fundamental areas of concern that we will need to addressed when designing, and starting our new bank. The following aspects will need to be worked out, in order to, launch a successful banking entity. We will need to define our **Executive Summary, Company Overview, Industry Analysis, Competitive Analysis,**

Technological Assessments, Customer Analysis, Products & Services, Marketing Campaigns, and Financials. These factors will be used to compose the business plan needed to start our Blockchain bank. Do not take these factors lightly because this is the most important part of your bank building process. It is during this first phase of the process that will determine the level of success for your Blockchain bank. Please feel free to start working out the particulars of your own Blockchain banking business plan as we proceed through this process. Each of your business plans will need to include operations, management, and an appendix. I will elaborate on most of the these aspects, but I will keep some of my cards close to my chest. But, nonetheless...let's get started on our global blockchain banking business plan.

○ **Executive Summary:**

RHGB (Rabbit Hole Global Bank) is an entirely new alternative to the failed legacy banking system. RHGB is based completely on Blockchain technology, and has an alternative design to both outdated traditional banks, and current Fintech structures. RHGB will use a multitude of Blockchain protocols to replace the departments of traditional legacy banking services, and products. RHGB's Blockchain-based structure is a hedge against the coming collapse of the global banking system. RHGB is also the future design of a global financial services platform that will replace the current concepts of financial institutions. RHGB's use of Blockchain's decentralized nature will enable the bank to operate in both established, and emerging global financial ecosystems such as B.R.I.C.S. without censorship by the ongoing geopolitical changes. Rabbit Hole Global Bank is an all-inclusive financial entity that seeks to bank the un-banked citizens around the globe. RHGB will provide ease of access to both foundational, and high-end financial products & services via Blockchain technology.

RHGB will open self-serve Blockchain financial centers called Rabbit Holes that will function as branches. These branches will have a grouping of BTM's each programmed to have a few investment choices that allows the investors to directly invest with the various Blockchain protocols. With an average of three tokens per BTM, and about 10 -

20 BTM's per Rabbit Hole, each individual will have ample investment opportunities to benefit from multiple Blockchain ecosystems.

We will lower the barrier for Blockchain adoption, and crypto usage by using the Rabbit Holes as learning centers. The Rabbit Holes will serve as an one-stop walk-in shop for investments, indices, market watch, and crypto news distribution outlets. Some Rabbit Holes will be super-centers where various Blockchain, and financial courses can be delivered to increase the financial acumen of the populations. This will be a crucial piece in developing new markets, and customer bases.

◦ **Overview Of The Company:**

Global Blockchain adoption is still relatively low, and is between 5% - 10% of the global population. Our goal is to launch in the top 15 -20 crypto adoption countries initially where crypto usage is a common practice. In these countries we will introduce a full suite of Blockchain financial services, and products to customers that are familiar with the basics of crypto usage. We will cultivate these bases, and elevate the crypto holders to full-scale high-end banking customers.

RHGB will build a vast network of BTM's (Blockchain Teller Machines) to serve three main purposes. The first is to provide an easy on-boarding method to increase the Blockchain adoption rate. Secondly, the BTM's will serve as our cash deposits mechanism. And lastly, the BTM's will provide a conduit for direct investment into the various Blockchain financial ecosystems.

I've decided like most Blockchain advocates that the RHGB will not be incorporated in the United States, and will not use any of it's legal wrappers. We are considering countries who choose not to be enemies of the people's wealth, and Blockchain technology. The UAE is a strong choice. The United Arab Emirates already has two financial institutions that are considered crypto-friendly banks. I considered the Marshall Islands because they were working on new a DAO recognition legislation. But, whoever is advising them has arranged the DAO registration, and the annual renewal process setup like a man-in-the-middle get rich scheme. Blockchain truly isn't compatible with any central authority.

But, banking licenses will be pursued in crypto friendly nations, plus there are various ways for a decentralized entity to own global assets.

RHGB is a web3 on-boarding platform that functions as a gateway to all of the Blockchain protocols that offer the same financial opportunities that outdated traditional banks provide. RHGB will not provide any account creation, nor custody services. RHGB's platform will function in a similar capacity as price indices. The difference is RHGB aggregates the protocols so individuals can take advantage of the decentralized financial opportunities that they feel benefits their needs. RHGB organizes the access to these Blockchain opportunities by removing the barriers of a learning curve, and the burden of years of research. Our platform is an one-stop shop to access Blockchain Banking services.

RHGB will be built upon a web3 structure using an Unstoppable Domains domain name, a web3 website that is localized, and web3 hosting on InterPlanetary File System (IPFS). The use of an Unstoppable Domains service guarantees that we own our domain name instead of renting a domain from a centralized domain service. The InterPlanetary File System hosting system will provide a censorship-free hosting method that will add a level of immutability to protect our content from third-party interference.

The Blockchain banking structure will be built using multiple DAO's, and the various Blockchain Decentralized applications (Dapps) that extend the functionality of the DAO's. Each DAO will use Safe(wallet) vaults for top notch security. DAO Safes have a long history of security audits, and a very robust community that is built around it's technology. In fact, most Dapps nowadays are designed to be implemented with the Safe treasury ecosystem in mind. A DAO will be created for each banking department, and the financial services that operate in that sector. The use of Dapps will open up a host of cross-chain financial opportunities. This will enable the bank to manage digital assets, and investments on several different blockchains to maximize overall value growth.

○ **Industry Analysis:**

The ability to maintain a clear understanding of all of the moving pieces in the rapidly growing Blockchain ecosystem is becoming increasingly difficult due to the growing rate of adoption, and innovation. From time to time we need to take a step back to gain a greater perspective of the industry, and any new trends that are occurring in the ecosystem. All investors should take a snapshot of the Blockchain markets at least once a quarter to stay up to date on any major changes in the industry. This will help you maintain a greater scope of operations by being updated on the trends, new solutions, and current challenges facing the crypto community. Changes occur on a daily basis, and new industry drivers are introduced every few months. And, one of the fastest growing sectors in Blockchain is the DeFi ecosystem. The growing number of new blockchains, new DeFi protocols, and new procedures to use these services can be overwhelming. Having a clear set of metrics that you can revisit will help with your periodical assessments of the industry. You should also have some reliable data sources to analyze your set of metrics.

One of the best ways to assess industry metrics is to cross-reference information from leading data aggregators tracking the activity in the industry. And, building a Blockchain bank will depend on the decentralized finance sector for the majority of its servicing, and products. Therefore, we need to take a look at the top DeFi aggregators, and the DeFi metrics that will help us understand the current state of the DeFi industry. Here are some DeFi metrics that you can start tracking. The following DeFi metrics will help you analyze the industry: **Total Value Locked (TVL), Trading Volume, Adoption Rates, Industry Trends.** I will be using DeFiLlama.com, and Dappradar.com as my reliable data aggregators for our DeFi metrics analysis. Once we assess the DeFi metrics then we will be able to make an informed decision about the future outlook for the industry.

The **total value locked** is one of the first metrics that can give you some insight into the current health of the DeFi industry. This is a fundamental metric that measures the total amount of valued assets that is locked into a specific Blockchain, protocol, or the industry as a whole. The TVL is also a good indicator of the level of trust, and the rate of growth in the DeFi sectors. Comparing the TVL every few months will give you a viewpoint of the amount of growth that has taken place over time, and any changes that has occurred during that time period. Currently in mid-2024, the DeFi market has around $170 billion dollars locked up in total value in various Blockchain protocols. The TVL will consist of the sum of total values of all of the individual sectors such as lending, stable coins, etc. This means you can calculate the TVL of a specific protocol, or a specific DeFi network, in addition to, the entire industry.

After looking at the TVL the next metric that I would take a look at is the **trading volume.** You should look at trading volume in terms of daily, and long-term trading. Comparing the trading volume will give you insight into the market activity, and the customer interest in the DeFi industry. You can derive a lot of information about consumer sentiments, and engagements from the trading volume metric data. You will be able to see what are the most popular protocols that are operating in the industry. Analyzing daily, monthly, and annual trading data will give you insight into the various trends, and changes occurring inside of the markets in real time. This is how you can identify where the majority of the liquidity is being employed, and where there is a need for additional liquidity. Currently, there is between $4.5 - $5 billion dollars being transacted on a daily basis. But, the metric keeps growing in tandem with the growing rate of adoption.

The **adoption rate** for the DeFi industry has experienced a varietyof change in the past few years. Currently, there are around five million users of various DeFi protocols, and services. There was just under eight million users three years ago during its peak. This drop might seem dramatic until you factor in the fact that there were higher security risks, and a few harmful computer hacks made the headline news a few years ago. There were some major losses that frighten a significant number of users away from the DeFi market for a little while. But, since then security audits has become a standard part of the industry, and the DeFi sector is regaining a lot of loss ground with Blockchain users. Plus, there is always a big wave of people that want to try out the newest Blockchain solutions. The metrics has shown that there is a steadily growing rate of adopting decentralizing finance solutions. That data shows that the Defi is expected to grow between 30% - 60% in the next few years. The overall growth of the crypto adoption rate will definitely contribute to the overall adoption rate of the DeFI markets. In 2024. you definitely have more serious investors in Blockchain than the earlier hobbyist adopters. Plus, the groundwork has been laid for institutional investing so you can expect a greater volume of investments, and exposure in the DeFi markets.

When you look at all of the trends over time you can get a sense of actual market patterns. These patterns will give you a view of the future outlook for the industry. The top trends in the DeFi market involves the investment of personal liquidity via various methods in the markets. Some of the ways that liquidity is being invested in DeFi is through yield farming, liquidity pools, and staking. Current trends show the growth, and security in the DeFi market is being propelled by the community members supplying liquidity to the Blockchains, the protocols, and the markets that they feel most comfortable using for their transactions. There are major opportunities to provide liquidity to DEX's, yield farming mechanisms, liquidity pools, Blockchains, and staking protocols. Every newly minted token, and new Blockchain networks will need immediate liquidity to become secure, and competitive in the market. So when the market grows to incorporate new solutions, and services then the needs for additional liquidity will grow as well.

∘ **Competitive Analysis:**

Rabbit Hole Global Banking platform is currently the only model being put forth as a total Blockchain Bank. The majority of the population stills has a conditioned legacy mentality that caters to custodialism. The future is a place where traditional banks do not exist, and most people are still in a place where they cannot totally conceive that reality. There is a significant quality difference between those engaging in passionate research, and the stale outcomes of paid, or technical research. Our vision is organic, and you can see it growing at the cellular level. You will not find to many people who has the passion to succeed legacy banking, and have a passion for implementing cutting edge Blockchain technology in the financial sectors. In fact, most people operating in the crypto ecosystems have only been involved for a hand full of years. I have been following Blockchain shortly after the original white paper was posted. I've seen Blockchain grow from the seeds that were planted in the sub-cultures of the web to presidential campaign planks, and global regulations. We've had our 10,000 hours in Blockchain long before governmental agencies created task force teams to monitor the deep web. We knew crypto was going to be legal when the first deep web market had been taken down, and the feds ended up with 13% of all of the Bitcoin in circulation in their forfeiture coffers. I remember earlier attempts to introduce other forms of digital currencies such as E-Gold, DigiCash, and the Liberty Reserve. I've had a vantage point for a long time that gives me a clear view of the entire field, and we have built our vision by surveying the entire landscape not just the limited screen view. While most people still thinks Blockchain is a little cute baby, we understand that this technology is going through puberty, and we're preparing it for adulthood. Those people who have a strong acumen dealing with legacy financial instruments have benefited too much to bite the hand that feeds them. They have no way to conceive a reality without their outdated pieces on armor. These are the people who never experienced a financial system that has discriminated against them, nor have they witnessed any economic exclusion. Only certain individuals have the balance that's needed to grasp the vision of a system of global financial platforms accessible for all people to use. Very few people have given any thought about a life being built outside of the legacy banking system. A person who has studied the history of money, and is operating on the far reaches of financial innovation is in an unique position to create a paradigm shift. Especially when you have an indigenous global world view that keeps your pulse on the needs of the people all around the globe.

My historical understandings of monetary policies, my genetic journey through the ills of society, and my family tradition of community stewardship has placed me in the right place at the right time to organize the pieces of change. Having an organizing dynamic will be essential to the platform's success. Having a background of community organizing, and campaign coordination will bode well for building organic community support. This skill set will be needed to help change the minds of those who do not understand the benefits of a non-custodial economy. The ability to market a completely new way of living, and bring that lifestyle to the global market is a declaration that we have gladly accepted. Being the first one to introduce a new way is not easy until you show others how it can be done. This is what we are doing with the Rabbit Hole Global Banking platform. Our competitive edge is there is no competition when you are the first to make it happen. There are only those who follow you on the path that you've laid out.

° **Technological Assessments:**

To build out our Blockchain banking platform you will need about a dozen different pieces of technology to start creating the actual structure. Then you can add several additional aspects to your platform to customize it to the community that you will be serving. The customization of each platform will be essential to success for each market, and business model.

    i. Web3 platform-
   ii. InterPlanetary File System-
  iii. Unstoppable Domains-
  iv. Blockchain Social Platforms-
   v. NFT Gated Community-
  vi. Decentralized Autonomous Organizations-
 vii. Decentralized Applications-
viii. Smart Wallets-
  ix. Decentralized Finance Protocols-
   x. Localization Software-
  xi. Tokenomics-
 xii. Blockchain Teller Machines-

Having well experienced Blockchain programming developers on the payroll will open our Blockchain bank up to an additional world of benefits, and opportunities beyond supporting established protocols. Eventually, we will need to develop our own Blockchain Teller Machines that can be programmed with several DeFi protocols for direct investments with fiat notes. Having multiple BTM's with multiple investing options in various locations will constitute a Rabbit Hole. A rabbit hole will be the new branches of bankless Blockchain banking.

◦ **Customer Analysis:**

There are several different customer pools that will funnel into Blockchain banking platform. The first group is DeFi users who are already using certain aspects of these Blockchain services. This group will be upgraded to included additional services via platform. The second group of customers will come from the people who use crypto, but have never used DeFi because of the lack of understanding. Removing the barrier of difficulty will add in this process. The majority customer pool will be newly on-boarded individuals who are introduced directly into the growing pattern of the DeFi market via our web3 platform. Some of these people will be from the legacy banking sector, and some will be the ones who were excluded from the legacy banking system. The next pool of customers will be financial investors looking for the next big thing. We will benefit, and grow in tandem with the global adoption of Blockchain.

◦ **Products & Services:**

The initial products, and services will come from organizing numerous Blockchain protocols that provide access to a variety of financial instruments. These are the type of Blockchain protocols that will cater to the following financial services:

- Lending & Borrowing
- Liquidity Investing
- Insurance
- Markets (capital, prediction, etc.)
- Token Launching
- Real World Assets
- Wealth Management
- Store of Value
- Consultation Services

We will also offer an online learning community to re-enforce financial acumen, and platform usage. RHGB will be a web3 portal for a financial membership community. The creation of Rabbit Holes as a self-custody one-stop Blockchain investing locations via BTM's hubs.

◦ **Marketing Campaigns:**

All marketing campaigns are dependent upon the original purpose of the business. Understanding the purpose of the business will give you a set of business goals. Once you have a set of goals then you can begin to set the preferences, and priorities of achieving each of those goals. You can determine the order of your goals by the business needs, or by the most efficient way to achieve those goals. When working towards achieving your goals you need to develop a means for accomplishing each goal. This is very important because to build a successful business you'll need to build your entire business structure to support, and achieve the goals of the business. To achieve this task you will need to identify the markets for each of the business goals. After you have matched the markets with your business goals then you can start brainstorming on ways to work those markets to your benefit. This is how you should be creating your customer funnel. Start looking at the best methods, activities, and channels to funnel people from each market into your business structure. Marketing will be determined by the character of each of those markets. The character of the markets is the collective character of the people who make up those markets. How do these people think, and act? Knowing these answers will help you identify the best channels of communication, and the type of activities that the market will respond to in a positive manner. Start generating some activities, and re-enforce those activities with methods of support. Methods of support can involve collecting market data like a contact, and emailing list for future marketing steps. Other methods of support could involve smoothly delivering supporters to your gated payment system. Possibly it could mean providing additional marketing support for customer retention. Marketing is a process that needs to be refined on an ongoing basis. Review your results with the original business goals to see what changes can bring improvement to your marketing efforts. Now begin stacking, and ordering your marketing activities to form the management of your marketing campaigns. Track the metrics, and look at the data from your marketing methods. Determine the frequency, and cycles of the most successful activities. Use this information to create campaigns that will increase the value of your products, and services. Refine this process to maximize the funneling of your targeted markets into your business structure that is supporting your business goals. Your business goals are supported by your business structure, and your business structure is supported by marketing campaigns that support the markets associated with your original business goals.

∘ Financials:

This is a very personal aspect of starting up a business, and each situation will have a different set of numbers, and needs. Working out the financials of a business is layered upon the process of working out your personal finances. Starting a business means creating an additional business budget plus maintain your personal budgeting for everyday life. You will need a budget before, during, and after you start your business. Each individual will have their own set of needs, and funding. Your budget can be altered according to the skill sets available to you, and your team. Why pay for services that you can provide for yourself? Creating a budget is a process of monetizing all of your brainstorming, and business planning to position yourself for successful business venture. Just start with a basic budget, and then expand your thought process to include the future growth of the business. Think about all of your costs needed to just plan your business. Then move to what funding is needed to start your setup. Map out what will be needed to just maintain the business on a monthly basis. This will be your overhead costs that is included in the business budget. How much funding do you actually need versus the amount of funding that you want for your business? How much will it costs you to seek additional funding? Are you launching a token? What will the entire process cost you? Do you need a travel budget to get to various pitch events for funding?

What about websites, sub-contractors, and branding merchandise with your logo. The more thought that you put into your business planning, the more detailed your financial reports, and your budgets will be for your business. This is how you can begin to build out your marketing campaigns.

CHAPTER V

Security & Banking-
Blockchain Banking Vaults

Throughout history the pervasive logic, and fundamental reasoning for the creation of banks is the heighten perception of banks to provide elevated security measures for your value's protection. The belief that banks can protect your valuables, and financial wealth a lot better than you can is the foundational premise of our modern banking system. Somehow people have accepted the belief of external security, and third-party custodianship for personal wealth management. It seems like an easy argument to win telling people that they should be responsible for their own financial well being. But, oddly enough this is far from the truth because the majority of people are conditioned to surrender their accumulated wealth to unknown financial managers. Somehow this has become the normalized practice of the accepted financial growth process in modern societies. Are these unknown financial managers focused on your well being, or are they dedicating their time to achieving their alternative goals. So, what is the true status of security within the legacy banking system. And, does Blockchain provide any positive benefits in the areas of financial security. If so, what are the advantages of Blockchain security in promoting both the growth of value, and the protection of yours assets.

One of the most important benefits that Blockchain gives individuals is personal control, and financial freedom. Blockchain guarantees these benefits throughout your entire decision making process. Blockchain gives a person the ability to have control of the complete financial transaction from decision to verification. This gives you the ability to benefit from all financial opportunities in real time instead of missing out due to delays in the legacy banking process. Account creation, and account management has more flexibility, and more user options when using Blockchain technology. You can create a new account immediately for whatever financial need that arises. You can create accounts by tasks, by transaction categories, by separate invoices, or just to have some extra accounts available for future needs. This reduces the security risk of having multiple transactions running through the same couple of accounts. This protects you from losing everything by having your main bank account compromised by cyber crime. In legacy banking systems the new account creation procedure is a very restrictive process. To create a new legacy bank account a person will need to re-submit more personal, and private information. Then, they will need to wait during the review process to see if they are given permission to continue the account opening process. A secondary confirmation procedure needs to be performed to insure the information submitted was correct, and accurate. Finally, if the individual is allowed to open a new account then they will need to fund that new account before it can receive a transaction. This burdensome process often leads to people trying to manage their entire financial lives through one, or two accounts. This model has shown us repeatedly that one compromised account can destroy a person's entire well being.

The permanent accessibility of accurate financial records through Blockchain for both past, and present transactions is another benefit of using Blockchain technology. These records are immediately accessible on the Blockchain at any given time regardless of time, place, or situation. Information stored on the Blockchain will always remain on the Blockchain with open access to everyone who might need access to those records. Plus, Blockchain wallets come with additional security features that regular banks do not provide people, an account recovery feature. The wallet recovery process gives you the added layer of security by giving you a valuable recovery method to regain access to your crypto in case you lose your wallet, or mobile device. This method comes in the form of Recovery Words that are generated during your initial wallet account creation process. When a person loses their crypto wallet the can just get a new wallet, and use their recovery words to regain access to all of the accounts associated with that wallet. Recovery words are an extra set of keys to access your digital assets, in addition to, passwords, and pin numbers. It is through these same recovery words that access to generational wealth can be passed down from generation to generation.

The security benefits of Blockchain financial models are secured by more than just security features. A major aspect of securing financial goals is access to financial opportunities. Using Blockchain gives you the opportunity to make financial decisions 24 hours a day/365 days a year. Blockchain does not close on holidays, and weekends. Blockchain doesn't shut down because major markets are in panic. No politician, or CEO can order Blockchain to close until further notice. Geo-politics does not have borders in the world of Blockchain. The most secured mode for Blockchain mobile access is to have a hardware wallet that connects to a secured dedicated financial mobile device with internet access. A hardware wallet connects to a wallet app to view, send, and confirm transactions. This is how we secure our value with Blockchain technology. There are different types of wallets, but the most secured type of wallet is a hardware wallet. And, the most secured Blockchain Dapp is the Safe Global wallet vaults. Safe Dapp provides the best in security features for Blockchain wallet protection. Safe is probably the most audited, and accepted Dapp operating in the wallet development, and web3 space. Safe has basically become the standard for wallet, and vault design for securing blockchain transactions. An entire community of decentralized application developers have built various ecosystems around the Safe Dapp capabilities. The majority of DAO's use Safe

Vaults to manage the DAO's finances, and governance decisions. In fact, there is a whole community of Dapp developers who have built a host of capabilities for the Blockchain ecosystem using Safe's security features. You can start a DAO, a Blockchain Bank, and a new Career with a hardware wallet, a Safe wallet Dapp, and a connected mobile device.

The Blockchain community is making tremendous strides in the area of financial security, and value protection. Blockchain's security industry is one of the fastest growing sectors in the crypto space. Developers in the Blockchain security space are providing the essentials services, and products for the rapid rise of new Blockchain solutions. This response stems from past security failures that have left people with financial hardships, and losses in the early Blockchain era. These past experiences has created a code auditing, and security posture culture within the Blockchain community. Now, you will find Blockchain security firms, and tools mushrooming around each and every layer in the Blockchain's architecture. The first line of defense is code review. Making sure Blockchain solutions have been audited by qualified teams are crucial to any security framework. A weak code with bugs in it can threaten the data layers, protocol layers, and network layers. How the data is processed will affect how the protocols will perform. These protocols provide the governance needed for data interactions between the networks to communicate with each other. So, to secure the entire Blockchain architecture professional firms, and tools have been developed to audit security at each layer in the Blockchain architecture. Having an abundance of specialized expertise applied to each layer in the entire application process creates scalable opportunities.

Some of the common security features being addressed by Blockchain security firms are smart contract auditing, transaction processing, network analysis, and securing web3 platforms. Now, we have tools for wallet protection, infrastructure management, threat assessment, secured project development, and network monitoring. Some firms secure Blockchain networks, and some handle risk mitigation. There is also a growing bug bounty movement growing in the Blockchain developers community. It is common practice, and a required process to have regular audits for web3 projects, and new Dapps. There are a few aspects of a Blockchain project that make it a good investment. Of course consistent organic growth is the obvious factor, but having a healthy synergy of security and trust is also essential. The trust factor stems from a proven record of maintaining a secure network.

When securing technology, or reviewing the security posture of various technologies, the best place to begin is at the code level. In the Blockchain ecosystems smart contracts are the most vital pieces of computer coding that needs to be secured. This is paramount for both Blockchain developers, and users of these protocols. Here are a few solution providers for smart contract auditing. The list is not ordered by the best, or preferred choices nor affiliation. The list is just some basic starting points to begin your security research while doing your due diligence. As an exercise take a look at some of the smart contract auditing portfolios of these companies listed below. Each company might have a different methodology in the way they approach their auditing process. Checkout what type of projects they have audited in the past, and how many types of projects the company has audited to date. Were there any exploitation incidences post auditing service that raises red flags about their auditing capabilities? Each of the companies below have conducted hundreds of Blockchain smart contract audits to date:

- SlowMist
- Hacken
- Trail Of Bits
- Halborn
- Armors

Once you are convinced that the Blockchain protocol has a secured coding audit process then your security outlook should focus on wallet security. Where you store your value is just as important as the locations that you acquire your value. First of all, your wallet needs to be **NON-CUSTODIAL**. This means nobody holds, or controls your digital assets but you. You should have full control of every aspect of the financial transaction from decision-making to transfers. Do not get fooled by centralized companies like Metamask/Consensys pretending to be non-custodial for profit gains, but have a history of seizing the crypto wealth of its users. A lot of Asians had their wealth confiscated for no reason but their ethnic demographic. There is a growing number of legacy-based entities masquerading as Blockchain solutions. This is why doing your own research is culture for securing your financial success. In terms of wallet security the best options are always hardware wallets. Hardware wallets are similar to physical USB drives, but designed to safeguard Blockchain transactions. These types of wallets remain offline until you are ready to give your permission to authorize any financial transactions. Here are a few starting points to begin your wallet security research:

- Ledger Nano Wallet
- Trezor Wallet
- SafePal Wallet
- Atomic Wallet

These hardware wallets connect to both Dapps (decentralized applications), and regular apps to facilitate online transactions. The applications function as user-interfaces with additional wallet capabilities. This means that the online applications needs to be secure, in order to, secure the entire wallet storage process. One of the most trusted multisig smart account management platforms is Safe Global. Their Safe Core vault technology has garnered a very robust community of developers that have built an entire ecosystem of Dapps around the integration of the Safe Global platform. You can manage the entire treasury of a DAO with the Safe Global Dapp. You can build a Decentralize Autonomous

Organization with a hardware wallet, the Safe Global Dapps, and a DAO software pack without any coding knowledge.

As an investor in Blockchain technology you will be interacting with several different decentralized networks. It would be best to start looking at the best ways to insure the networks that you use are being secured properly. There are multiple methodologies for securing Blockchain networks including tools, and private companies. One of the best decentralized methods for securing Blockchain networks is Forta. Forta is a tool that is used to monitor malicious activity in wallets, DeFi, and Blockchain bridges. Forta is credited for successfully detecting million dollar exploits in major DEX's, and DeFi protocols. A decentralized security solution for decentralized networks, and decentralized applications is true to Blockchain's nature.

Another important Blockchain security tool is the ERC-20 Verifier tool. This is a tool that checks ERC-20 tokens for compliance. This security tool makes sure smart contracts comply with the standard ERC-20 token formats. The ERC-20 Verifier tool is used by both general users, and Blockchain developers to confirm smart contract compliance requirements. This compliance tool is a product of the OpenZeppelin security platform. There are hundreds of various Blockchain security tools, and companies. Everyday more solutions are being developed for the community. Make a habit from time to time to perform a basic search of the current top Blockchain security solutions just to stay updated on the security options available to you during your investment journey.

Having a **Self-custodianship Mindset** is actually the very first layer in Blockchain security framework. It is true that this mindset requires more hands on attention than your current legacy banking practices. This is only because the legacy banking system fosters a financial slave mentality that convinces people that they will always be incapable of managing their personal freedoms. In Blockchain there is more hands on responsibility, but there is also more benefits, and more financial opportunities available to the user. Building a financial due diligence culture is essential to achieving financial success. Changing your financial mindset is the first step in securing your financial destiny. Those people who continue to believe that corporations have your personal interests above their own interests, their own profits, and their own corporate destiny are destined to fail in achieving financial success. Nobody owes you anything...including all of your money that you give away to the banks. Once you give your wealth to the banks that money becomes the bank's wealth. You have to borrow pieces of that amount back on an ongoing basis. This is why you have to pay them fees to get "your money" back from the banks. You don't tell banks what to do with your money. The banks tell you what to do with that money. Whenever you withdrawal money from a bank it is consensual agreement, and not a "guarantee" of a secured service. The truth is legacy banks cannot secure your value, not even for their own benefits. A quick look at legacy banking's security posture will reveal that they are incapable of maintaining the first rationale for a bank, Security.

There are some basic components of legacy banking security that people will need to focus on, in order to, secure their financial well-being. These factors are physical protection, virtual protection, and institutional protection of your assets, and value. Physical protection involves preventing the physical theft of your valuables, and assets from within the banking institution's store front. It is based on securing the actual perimeter of the bricks, and mortar where your wealth is stored. Literally everyday multiple banks are being robbed by every type of person in society from children to senior citizens. We are talking about thousands of bank robberies per year. And, physically robbing banks has become a second-rate theft compared to the increased theft from banks via online platforms. With all of the technology improvements in legacy banking the banks still have an ongoing increase of thefts via their virtual platforms every year for the past couple of decades. When you take a good look a the top banking data breaches you will notice that almost every person in this nation has had both their value, and their personal information compromised. Not only did banks fail to keep your wealth save, but they double-down on a failed security posture, and failed to keep all of your personal information secured. This is the same information that banks claimed they needed to safeguard your valuables just for you in the first place. Now, you have cyber-criminals that can monetize your personal information in ways that you never knew existed, in addition to, taking your funds. But, there is an even greater danger of financial loss than physical, and virtual insecurity. That is the lack of protection from institutional theft, and the confiscation of money, assets, and wealth. Hundreds of banks are failing, and even more banks are in the process of collapsing. Banks are private businesses, and any private business can file bankruptcy, or decide to closed its doors, and cease business operations. At anytime a bank can decide to stop doing business with you without any face to face explanation. Have you ever used your ATM card, and it didn't work? Did the bank call you, or did you have to call the bank to ask why you could not access your money? Have you ever had one of your accounts closed without your consent? Maybe for "security purposes", or "for your safety". Maybe you should take a little time to research how many people around the globe cannot access their own wealth due to banking crisis spreading from country to country. Maybe you are totally fine with receiving pennies on the dollar, or limited allowance of withdrawals. Maybe you just don't care if others just take all of your money, assets, and wealth from you whenever they decide it's in their best interest. Well, I care about my financial security enough to learn new ways to manage my wealth. And, Blockchain has given me the tools I need to secure my financial destiny.

So, if you truly want financial security for your family, and yourself then you must adopt a self-custody mindset. Then, you need to create a culture of financial research to strengthen your financial acumen. The next step is to adopt the technology, and tools that will assist you in achieving your financial security goals. You are the only person that is going to focus all of their energy, and time on researching, building, and securing your financial future. If you don't believe me...then make a long list of all of the people who will dedicate their lives to securing your financial future. Then, contact each person, and ask them where are you at in the process, in regards to, achieving your financial destiny.

CHAPTER VI

Money Creation-Tokenomics & Value Creation

Anyone who says money is the root of all evil obviously doesn't know a damn thing about the history of evil, nor the history of money. These people are human-parrots who believe parrots are brilliant because the birdbrain can repeat a few words from an idiot. The historical timeline of evil predates the creation of money by countless light years. Modern humans are still attempting to document our origins accurately. Therefore, claiming to know the origins of evil itself is still a work in progress.

On the contrary, the origins of money in all of its various forms are well documented throughout human history. The various methods of measuring, and exchanging value in our daily lives is an essential function of survival. We structure our lives through the organization of values. This is true for moral principles , as well as, physical resources. Finding, and accessing value is a major driving force in our lives. Evil is the exact opposite of this process of living. When people do not want other humans to feel valuable, and deem their lives as worthless...this is true evil. "Evil" is literally "Live" spelled backwards, and so is the process. I'm writing this book to add value to your lives. So you can free yourself from a financial system that is controlled by very evil people. Do not get it confused about the nature of value because confusion is the tool of the devil's advocates. And, nothing in modern society is more confusing than our dysfunctional financial systems.

The meaning of Life, and Love may be the only human concepts that holds more mystery than Money. It's bizarre behavior to see humans spend their entire lives interacting with an abstract concept that they don't even understand. I equate this behavior to being on suicide watch while eating your favorite popcorn. Society loves to entertain its members by interacting with mysterious concepts. The mystery fades away we you realize the difference between Value, and Money. People try to convince you that they can add money to your life, but in reality all they can do is add some value. For instance, having self-worth is an essential value for achieving financial success. No one can give you self-worth because this value resides within the individual. Money is just a conceptualized tool that we've invented to measure the acquired, and unattainable values in our lives. Constantly adding value to our lives is considered 'good living'. The consistent loss of value in our life is deem as 'wasting life'. The term that we use to describe this negative type of process is called poverty. Poverty has negative affects on the living process. Wealth has a positive effect on our lives. Wealth is the process of consistently adding additional value to your life. Money is just a method of measuring your deeds, and efforts.

It makes no sense that any modern society can have a phenomena called Generational Poverty. Generations are made up of hundreds of people who create multiples of each person. Each of these individuals has a set of values, and spends most of their lives striving to acquire additional value to add to their lives. So ask yourselves how people can spend everyday of their lives working to provide a better life for their families, and still fall short of financial success. How does the majority of people on the planet know how to transfer poverty to the next generation, but have no insight on how to transfer their accumulated value to their offspring? We are born on a planet of abundance. You cannot dispute the fact that there is enough for everyone. Any skeptic only needs to look at the fact that those people who exploit the worlds resources have never ran out of resources to exploit in any part of our human history. The ongoing transference of poverty is not a natural process because natural processes facilitate humanity. They do not restrict a human's ability to live, and exist in a world of abundance. What this means is generational poverty is a man-made process that is well-maintained throughout a person's entire life. If there are groups of people working everyday to keep you financially ignorant then "We The People" need to work on building up our financial acumen everyday.

As mentioned earlier, the long history of money is well documented so what can we derive from this accumulate history. Over time through trial, and error we have established certain criteria that must be met, in order to, be considered as an acceptable form of the term we coined as 'Money'. Money is just a tool to measure various units of value. It's something that we can use as a medium of exchange. A way for humans to transfer value between each other as needed. In order to address both our small, and our larger exchanges, our money needs to be fungible in both small units, as well as, large units. Both the large, and small units need to be mobile, or portable. And, money needs to be able to last a long time not just for today's needs. This means money needs to be durable to last over time. Now, this time limit will vary according to the type of medium being used as money. For instance, a commodity-based form of money such as cocoa, or grain will have a different shelf life than a coinage-based money. Another major aspect of money is kind of abstract by nature. Humans need to provide consent, belief, and trust into their form of money. Here lies the esoteric origins for the mystery of money. This is where the crossroads for financial freedom, and economic slavery resides. The belief, trust, and consent of money is the mystery of all financial destinies. Both individual, and collective economic success hinges on this abstract concept of money.

Please remember this truth (...belief, trust, and consent...) when moving forward in your lives.

Let's talk about "We The People...", and our current broken relationship to money creation. In the United States of America the constitution has a hand full of rights that defines the creation, management, and regulation of money. I will briefly speak on this relationship because of the U.S.'s power to set monetary policy on a global scale. Even though that monetary power is collapsing at a blinding rate. This is what the constitution says about money:

- *Article I, Section 8, Clause 5- Congress shall have power to coin money, regulate the value thereof, and of foreign coin, and fix the standard of weights and measures*
- *Article I, Section 10, Clause 1- No state shall coin money, emit bills of credit, or make any thing but gold and silver coin a tender in payment of debts*
- *Article I, Section 9, Clause 7- No money shall be drawn from the Treasury, but in consequence of appropriations made by law*
- *Article I, Section 8, Clause 2- Congress shall have power to borrow money on the credit of the United States*
- *Article I, Section 8, Clause 6- Congress shall have power to provide for the punishment of counterfeiting the securities and current coin of the United States*

The constitution does not say that a group of a half-a-dozen greedy private bankers can have a secret meeting on Jekyll Island during the holiday season behind the backs of American citizens to steal the national constitutional rights, and future value of an entire nation. Nowhere in the constitution gives permission to a group of thieves to steal the nations gold out of the Treasury, and replace it with worthless pieces of paper. It doesn't say six people can decide to take the countries right to coin its own money from the nation so America will have to borrow its entire money supply at a compounded interest rate. Where in the constitution gives permission to the Federal Reserve to engineer a National Debt for future generations. Do not confuse the term federal with national. National refers to a nation's status. Federal, or federated just means a compact agreement of entities (private banks) forming an union. It is illegal for the Federal Reserve to replace America's real coinage with their own worthless fiat paper reserve notes. In fact, the constitution says that fiat notes (bills of credit) are illegal, and are strictly forbidden. The constitution even defines gold, and silver as the two metals that are legal to coin money in America. You can also find the exact measurement of the precious metal grains that are needed to coin one dollar.

One of the most forgotten concepts of modern money throughout history is that money needs to have a relationship with a precious metal. Let me be more specific about the physical composition of modern money. Real Money has always been coined. Money is supposed to be minted coins of precious metal backed up with organic economic activity. In the agricultural age money was organic products of high value backed up with organic activities. Even in these times money was being coined out of gold, silver, copper, or other metals by different cultures. The point is the term money has always meant physical coins of a limited supply of precious metals. This seems like science fiction in today's societies of fiat-based reserve notes. A quick look at America's M1 & M2 money supply charts will give you a graphic view of how dire the money creation crisis is to date. Looking at the graph you will see when the country was taken off of the gold standard, and when the feds just felt like they had no more obstacles to just print useless fiat notes at will.

According to the private federal reserve system the cost to print America's paper reserve notes was $931.4 million just shy of one billion dollars. And, every year the budget increases without constitutional enforcement. For an 115 years the United States of America has loss its right to coin its own money, and consented to an inflationary fiat note supply. The more fiat notes the federal reserve prints, and charges America for its monetary supply the less value each note will have on an ongoing basis. This illegal centralized authority over the citizen's money creation policy has proven itself to be a direct attack on the freedoms of the citizens. It is that time again when citizens must defend themselves against foreign, and domestic enemies who use financial weapons to destroy the freedom, and rights of the population.

Thanks to Blockchain 'We The People' have some solutions, and methods to address the money, and value creation crisis that has been forced upon us. This solution comes in the form of Tokenomics, and there are various methods to approach your financial goals through tokenization. Most people think of money as value, but the truth is money is more of a tool to measure value. The difference is money is incestuous, and value is promiscuous. We like to keep money in the family. But, with value you can give value to anyone. The more value that you provide for others the more valuable you become yourself. Accumulation of wealth is the process of acquiring wealth on an ongoing basis. Tokenization is a method of measuring, and creating value digitally using Blockchain technology. Tokens have several different functions. They can be a digital form of money, or a digital asset. Tokens can represent the right to access financial products, and services on certain Blockchains. Tokens can be used to denote membership in an Blockchain organization, and give that member voting rights in their governance process. Tokens are used to convert paper fiat notes into digital Blockchain units of value.

All tokens are not the same. It is important to understand that tokens support different ecosystems within the Blockchain space. And, each of these ecosystems have different functionalities therefore each of these tokens have different functionalities. Some token are fungible meaning they can be broken down into various measurement units such as cryptocurrencies. Then, you have other tokens that are non-fungible, and considered to have unique characteristics. These types of tokens are called NFT's (Non-fungible Tokens). Blockchain has an eco-system called GameFi which stands for Game Finance. In the GameFi financial model tokens are used as financial incentives, and are rewarded in the form of NFT's. This form of tokenization helps gamers to monetize both their playing time, and gaming assets such as accessories, land, or weapons. Tokenization gives Blockchain gaming the ability to let game players own, trade, and sell their in-game items on multiple markets (primary & secondary) as digital assets of value. You do not have to even play the game to acquire these digital assets as an invest-ment. In the MetaVerse ecosystem tokens are used as a form of virtual land ownership for developers, and businesses. So, if you wanted to invest in the future of Metaverse projects you can achieve these goals by acquiring the tokens associated with each project. In DeFi (Decen-tralized Finance) tokens are used to provide various Blockchain-based financial products such as lending & borrowing, market trading, and other financial services. Tokens are used to participate in these financial sectors, and different Blockchains ecosystems. These are just some of the ways that tokenization is used to create, mange, and transfer value in various Blockchain ecosystems.

These are some of the ways that tokenization fosters value creation in the world of Blockchain. Now, lets examine how the tokenization process can provide some solutions for the financial problems of legacy banking, and your personal financial needs. First, we need to address the problems of the inflationary fiat notes, the creation processes, and restricted access to financial support. Then, we can look at how you can begin your own token creation process to suit your own financial needs. Inflationary reserve notes lose additional value with every new infla-tionary note that is printed, and introduced into circulation. A method for the people to participate in the fiat policy making process to make changes to protect their value does not exist. In contrast to inflationary fiat notes, tokenization gives people the power to predetermine the char-acteristics, and mechanics of the token before it is available in public circulation. Blockchain tokenization gives you the ability to customize your financial instruments to meet both personal, and market demands. A token creator can determine the type of token, its utilities, the token supply, the rate of distribution, and the method of distribution among other factors.

To combat inflationary economics the majority of token designers will determine that the token supply will have a fixed limited amount. This is called a deflationary token supply when the amount of tokens initially created cannot be increased to devalue the monetary supply. This type of token mechanics creates an increase in token value as the token supply is acquired by token holders, and investors. This is one of the reasons that Bitcoin has such a bull market posture. Bitcoin was designed with a limited supply of 21 million units in its total supply. In 2020, 90% of all Bitcoins were already in circulation by the time the general public began realizing the value of holding Bitcoin tokens. This means that the tens of thousands of millionaires, and millions of people who had a desire to invest in Bitcoin had to compete for the last 10 %, or buy some Bitcoin from someone who already was a holder. The price of the token increases with both the demand, and the increased economic activity. Blockchain's decentralized network means that a token holder who was a child had the same ability to make a profit off of Bitcoin as an early Silicon Valley investors. This is true for similar deflationary token mechanics regardless of the Blockchain ecosystem. In addition to, creating a deflationary token you can decide whether you want the token to be fungible, or non-fungible. This decision will be based on the function, and utility of the desired token. The token design will determine whether it's a governance token, a transactional token, a utility token, a security token, or another form of a token.

Once the type of token has been decided with its utilities then the decision on its distribution process can be customized. A decision needs to be made about when the tokens will be distributed. You can even determine how the tokens will be introduced into circulation. Also, you can have a say in who these tokens are distributed to according to it's utility, and function. The recipients can be employees, developers, community members, or investors. Distribution can come in the form of a private sale, a public sale, a airdrop, through minting, or a rewards program. There are ample platforms currently available to assist in launching a new tokens with a simple intuitive user process. These platforms help with multiple aspects of the creation process from smart contract creation to launching the token. Once the token is launched then it is available in the global market 24 hours a day, and 7 days a week. This gives Blockchain markets an advantage over legacy banking markets that close every weekend, and on holidays. Now, that we understand the basics of tokenization we can start putting together a plan to launch our Blockchain investment journey to accomplish our financial goals.

Tokenization is a new solution that is available to the people to address our society's broken relationship with money. Governments of the world have forfeited the rights of the people to create their own money. The have become compromised, and cannot be relied upon to guarantee our financial well being. I suggest that you join the millions of people who are reclaiming the financial freedom through Blockchain tokenomics. Currently, it is the only suitable solution available to combat the fiat inflationary system that is collapsing at a record pace.

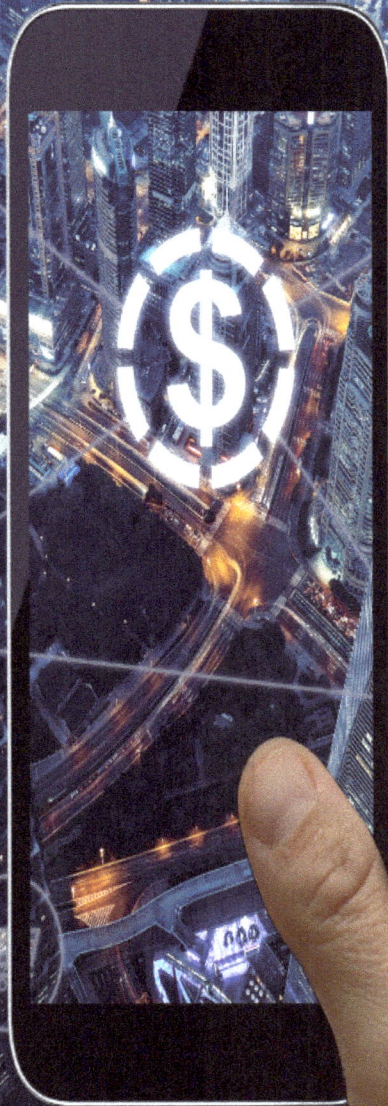
Banking At Your Fingertips

CHAPTER VII

Desigining A Blockchain Bank

From Concept To Structural Design

Rabbit Hole Global Bank: A New Blockchain Banking Concept

Now is the time that we can start dedicating some deep thought into the process of designing our Blockchain Bank. A lot of the initial input will depend on both the financial needs, and the financial goals of the banking business that we are seeking to build. Our goal is to build an open-sourced platform that has the full functionality of a global Blockchain bank. We aim to remove the learning barriers that prevent crypto adoption, and utilization of decentralized finance for the common citizen regardless of their location, or financial acumen. Our platform will be a web3 gateway that will become the one-stop shop for accessing various Blockchain-based banking opportunities. This platform will provide access to Blockchain protocols organized by service, and products. The Rabbit Hole will become a financial literacy community focused on fostering financial acumen, opening access to economic channels, and achieving financial goals.

This community will be built exclusively using Blockchain technology, and decentralized building blocks. Even the integrated social media platforms will be designed using Blockchain-based decentralized social media protocols. This will ensure the Rabbit Hole has a complete Blockchain-based foundation. The Rabbit Hole will be filled with learning modules, and courses that will promote the benefits of using Blockchain protocols in the pursuit of economic achievement. The use of Blockchain social media platforms will add multiple functions to the Rabbit Hole. Firstly, it will function as a friendly user-interface for the Blockchain DAO's that underline the platform's structure. Secondly, it will foster communication, professional networking, and community building around the financial platform. Another benefit of using web3 based social platforms is the ability of Rabbit Hole members to own all of the data that they generate on these social platforms. This will give Rabbit Hole members the benefits to monetize their social media activities. This includes their posts, comments, and other articles that they have generated in pursuit of their financial goals. This helps members create digital assets, and it help foster a self-custodial mindset which aligns with Blockchain's true nature. A self-custodial mindset is essential for accessing the platform because the Rabbit Hole is a non-custodial crypto platform. There **will not** be any account creation, nor account custodial services to be managed on behalf of others. All members will be responsible for their own financial destinies. Nobody will tell you that you can't do this, or that with your own resources, and financial decisions. Rabbit Hole will provide you with the choices, but the decisions are your responsibilities to make on your own.

Now that we've discussed the concept of our new bank, we can begin focusing on the services that we would like to provide as a Blockchain platform. There are five specific services that Rabbit Hole will provide via platform. Everything starts with learning the process, and the available opportunities. So the creation of a Blockchain learning space that promotes financial literacy via learning modules, and courses is essential. Rabbit Hole will be a learning academy attached to an self-investment platform. Providing support for both the learning process, and the investment journey will be crucial for collective success. Therefore, providing a community building, and support mechanism for the platform will be the next service provided. This will empower individuals to strive beyond their self-perceived obstacles, and limitations to achieve greater successes.

The main service provided will be an organized on-boarding investment platform. We are providing access to Blockchain protocols that are organized by banking service, or products to reduce the barrier of entry for Blockchain investing. The educational modules will reduced the learning curve for new members. The social community will provide support, and networking for the investors. But, the Rabbit Hole will provide access to all of the organized protocols that compromise modern banking services, and products. Here are some examples of the Blockchain protocols that provide banking functions in a decentralized manner. Some of these protocols are well established, and some of the protocols are fairly new to the ecosystem. Please, please, please **DO YOUR OWN RESEARCH (DYOR)** before investing into any financial transaction regardless if it is a Blockchain, or a legacy-based investment opportunity. All of the following Blockchain protocol listings are actual examples, and not recommendations. Each banking sector has multiple Blockchain protocols that can be used to facilitate that particular banking service. Let's break down some banking functions into Blockchain protocols to give you a clear view of how a Blockchain bank will take shape.

Banking Sectors:

Blockchain Protocol Name- Number (#) of Blockchains The Protocol Operates Upon:

Insurance:

- nexus mutual- 1 chain
- unslashed- 1 chain
- ease.org- 1 chain
- 3F mutual- 1 chain
- insurace- 4 chains
- sherlock- 1 chain
- cozy finance- 2 chains
- itrust finance- 1 chain
- ante finance- 8 chains
- cover protocol- 1 chain
- nayms- 1 chains
- neptune mutual- 3 chains
- ensuro- 1 chain
- nsure- 1 chain
- bridge mutual-3 chains
- tidal finance- 1 chain
- unoRe- 3 chains
- insure dao- 4 chains
- risk habor- 9 chains

Digital Money & Tokens Creation:

- uncx network- 7 chains
- pinksale- 6 chains
- team finance- 5 chains
- dxsale- 24 chains
- flokifi locker- 15 chains
- juicebox- 1 chain
- deeplock- 1 chain
- mint club- 10 chains
- vestige- 1 chain
- impossible finance- 7 chains
- thorstarter- 1 chain
- dexpad- 4 chains
- zkboost- 1 chain
- infinity pad- 3 chains
- chaingpt- 5 chains
- stealthpad- 1 chain
- aptoslaunch- 1 chain
- antex- 1 chain
- bounce finance- 1 chain
- westarter- 1 chain
- fuzion ignition- 1 chain
- kavastarter- 1 chain
- bchpad- 1 chain

- cardstarter- 1 chain
- decubate- 1 chain
- starterra- 1 chain
- merit circle- 1 chain
- polkastarter- 2 chains
- valkerie- 1 chain
- smartpad- 3 chains
- kommunitas- 2 chains
- seedify- 1chain
- kdlaunch- 1 chain
- axl inu- 2 chains
- ergopad- 1 chain
- velaspad- 1 chain
- uplift dao- 1 chain
- snark launch- 1 chain
- super launcher- 1 chain
- oxbull- 1 chain

Store of Value:

Staking Pools

- marinade native- 1 chain
- wemix.fi- 1 chain
- neopin staking- 2 chains
- abc pool- 1 chain
- filfi- 1 chain
- klayprortal- 1 chain
- thundercore staking- 1 chain
- tronnrg- 1 chain
- rakeoff- 1 chain
- fixes frc20 staking- 1 chain
- solana miner- 1 chain
- pls print- 1 chain
- quantum unit- 1 chain
- soarpls- 1 chain
- metald- 1 chain
- xbanking- 1 chain
- ultron staking hub- 1 chain
- unifi staking- 5 chains
- arpa staking- 1 chain
- ignore fud- 1 chain
- chainflip- 1 chain
- singularitynet agix staking- 1 chain
- gracy staking- 1 chain

Liquid Re-Staking Pools

- ether.fi stake- 1 chain
- renzo- 7 chains
- puffer finance- 1 chain
- kelp dao- 1 chain
- swell liquid restaking- 1 chain
- eigenpie- 1 chain
- bedrock unieth- 1 chain
- prime staked eth- 1 chain
- claystacked eth- 1 chain
- euclid finance- 1 chain
- restake finance- 1 chain
- inceptionlrt- 1 chain
- genesis lrt- 1 chain
- zero-g finance- 4 chains

Liquid Staking Pools

- lido- 5 chains
- rocket pool- 1 chin
- binanced staked eth- 2 chains
- mantle staked eth- 1 chain
- jito- 1 chain
- marinade liquid staking- 1 chain
- frax ether- 1 chain
- swell liquid staking- 1 chain
- stakestone- 1 chain
- coinbase wrapped staked eth- 1 chain
- stader- 6 chains
- blazestake- 1 chain
- stakewise- 2 chains
- benqi staked avax- 1 chain
- liquid collective- 1 chain
- glif- 1 chain
- tonstaker- 1 chain
- linear protocol- 1 chain
- stride- 14 chains
- meta pool- 3 chains
- synclub staked bnb- 1 chain
- amnis finance- 1 chain
- stacking dao- 1 chain
- marginfi lst- 1 chain
- bitfrost liquid staking- 5 chains

Credit:

- solv funds- 6 chains
- maple- 2 chains
- wildcat protocol- 1 chain
- clearpool- 5 chains
- union protocol- 3 chains
- ribbon lend- 1 chain
- atlendis- 1 chain
- truefi- 1 chain
- dAMM finance- 1 chain

Debt/CDP:

- makerdao-
- juststables-
- liqutiy-
- helio protocol-
- crvUSD-
- thorchain lending-
- abracadabra- 8 chains
- overnight finance- 8 chains
- indigo-
- liquid loans-
- orby network-
- pando leaf-
- sDAI-
- inverse finance-
- prosma finance-
- kava mint-
- fx protocol-
- thala cdp-
- sovryn zero-
- angle- 8 chains
- parallel protocol- 3 chains
- qidao- 14 chains
- lybra finance-
- powercity earn protocol-
- parrot protocol-
- usk-
- gravita protocol- 7 chain

Lending & Borrowing:

- aave- 12 chains
- JustLend- 1 chain
- spark- 2 chains
- compound finance- 5 chains
- venus- 3 chains
- morpho- 1 chain
- kamino lend- 1 chain
- marginfi lending- 1 chain
- solend- 1 chain
- benqi lending- 1 chain
- zero lend- 5 chains
- fraxlend- 1 chain
- radiant- 3 chains
- burrow- 1 chain
- orbit protocol- 1 chain
- silo- 2 chains
- avalon- 1 chain
- fluid- 1 chain
- layer bank- 6 chains
- kinza finance- 2 chains
- nostra- 1 chain

Liquidity:

- uniswap- 16 chains
- pancake swap- 9 chains
- curve- 13 chains
- balancer- 7 chains
- aerodrome- 1 chains
- raydium- 1 chains
- sun- 1 chain
- sushi- 35 chains
- thruster- 1 chain
- pulse x- 1 chain
- thorchain- 9 chains
- vvs- 1 chain
- meteora- 1 chain
- katana dex- 1 chain
- orca- 1 chain
- velodrome-1 chain
- osmosis dex- 1 chain
- sanctum infinity- 1 chain
- quickswap- 6 chains

Yield Farming:

- Yearn Finance- 6 chains
- Beefy- 27 chains
- Sommelier- 3 chains
- Badger DAO- 5 chains
- Yield Yak Aggregator- 2 chains
- Vesper- 4 chains
- AcryptoS- 17 chains
- Spool Protocol- 2 chains
- Harvest Finance- 4 chains
- Idle- 4 chains
- Pickle- 12 chains
- Autofarm- 20 chains
- Reaper Farm- 4 chains
- Matrix Farms- 6 chains
- Amulet V2- 4 chains

Payments:

- lightning network- 1 chain
- flexa- 1 chain
- stablier- 10 chains
- llamapay- 10 chains
- superfluid- 6 chains
- sushi furo- 11 chains
- roketo- 1 chain
- streamflow- 5 chains
- pulsar money- 1 chains
- parallel pokadot daofi- 2 chains
- ajira pay finance- 3 chains
- kivach- 1 chain
- coindrip- 1 chain
- telcoin- 1 chain
- retreeb- 1 chain

Markets:

Capital Markets/Options:

- thetanuts finance- 11 chains
- deri- 8 chains
- lyra- 3 chains
- hegic- 2 chain

- opyn- 3 chains
- dual finance- 1 chain
- cega- 3 chains
- opium- 4 chains
- premia- 5 chains
- psyoptions- 1 chain
- stryke- 4 chains
- smilee finance- 1 chain
- ribbon earn- 1 chain
- option dance- 1 chain
- zomma protocol- 1 chin
- siren- 3 chains
- moby- 1 chain
- rysk finance- 1 chain
- buffer finance- 2 chains

Prediction markets:

- polymarket- 1 chain
- azuro- 3 chains
- gnosis protocol v1- 1 chain
- augur- 1 chain
- etherflip- 1 chain
- winr protocol- 1 chain
- thales- 6 chains
- zethr- 1 chain
- zkasino- 4 cahins
- prdt- 4 chains
- ferdyflip – 3 chains
- megamoon- 1 chain
- bethash- 1 chain
- sportbet- 1 chain
- eesee- 1 chain
- whale game- 1 chain
- polkamarkets- 3 chains
- entropyfi- 1 chain
- betfolio- 1 chain
- prophet- 1 chain
- deeppdao- 1 chain
- catsluck- 1 chain
- nexter- 2 chains
- sagebet- 1 chain
- luckychip- 1 chain
- debets- 2 chains
- crypto lottery- 6 chains

- dropcopy- 1 chain
- lunafi- 1 chain
- fliperino- 1 chain
- loterra- 1 chain
- sunrise gamling- 1 chain
- totemfi- 1 chain

Leverage Trading:

- gearbox- 2 chains
- juice finance- 1 chain
- extra finance- 2 chains
- homora v2- 5 chains
- deltaprime- 2 chains
- alpaca leveraged- 2 chains
- airpuff- 4 chains
- bow leverage- 1 chain
- stella- 1 chain
- archi finance- 1 chain
- kleva farm- 1 chain
- factor leverage vault- 1 chain
- arcadia finance- 3 chains
- single finance- 2 chains
- steadefi- 2 chains
- archimedes finance- 1 chain
- sentiment- 1 chain
- zkfox- 1 chain
- jewelswap leverage- 1 chain
- vivaleva protocol- 1 chain
- bao baskets- 1 chain
- ithil- 1 chain
- zenith- 1 chain
- jbc.finance- 1 chain
- perseid finance-1chain
- pacman- 1 chain

Derivatives:

- gmx- 2 chains
- dydx- 2 chains
- hyperliquid perp- 1 chain
- jupiter perpetual exch- 1 chain
- drift- 1 chain
- apex protocol- 7 chains

- vertex- 1 chain
- parcl v3- 1 chain
- aevo- 3 chains
- synfutures- 5 chains
- apx finance- 7 chains
- mux protocol- 5 chains
- 01- 1 chain
- particle LAMM- 1 chain
- blast futures- 1 chain
- blitz- 1 chain
- hmx- 3 chains
- bluefin- 2 chains
- gains network- 2 chains
- kiloex- 3 chains
- sythetix v3- 1 chain
- umaproject.org

Real World Assets:

- ethena- 1 chain
- maker rwa- 1 chain
- stUSDT- 2 chains
- ondo- 5 chains
- realt tokens- 1 chain
- superstate- 1 chain
- mountain protocol- 5 chains
- hashnote USYC- 2 chains
- tangible rwa- 5 chains
- matrixdock- 1 chain
- maple rwa- 1 chain
- backedfi- 8 chains
- lofty- 1 chain
- openeden t bills- 2 chains
- aktionariat- 2 chains
- fortunafi- 3 chains
- meld gold- 1 chain
- solv protocol- 6 chains
- digift- 1 chain
- hiyield- 2 chains
- bloom- 1 chain
- landx finance- 1 chain
- toucan- 3 chain
- binaryx platform- 1 chain
- cache.gold- 1 chain
- gold dao- 1 chain

- kuma protocol- 3 chain
- anzen finance- 1 chain
- danogo- 1 chain
- frigg.eco- 1 chain
- sailing protocol- 1 chain
- sweep- 7 chains

You can also incorporate a form of cash deposits into your Blockchain Banking operation with the use of BTM's (Blockchain Teller Machines). BTM's are ATM's that convert fiat reserve notes into crypto currencies , and tokens. This is an extremely simple process in which an individual inserts cash into the BTM, scans their crypto wallet address, and then receives the stated amount of tokens via Blockchain to their digital wallet address. You can use a BTM to sell your crypto holdings for cash using a similar process. Most BTM's can be programmed to dispense two, or three different types of tokens. The Rabbit Hole format will use a network of BTM's that are programmed with the various Blockchain banking investment tokens grouped together by services. This would solve multiple issues including fiat conversion, investment barrier removal, and providing a simple user-interface among other benefits. A Rabbit Hole will be a place similar to a bank branch, but self-custodial in its services. Rabbit Hole members will have access to an on-boarding platform where they can take their cash, and deposit into the various BTM's to immediately diversify their Blockchain investment portfolios. A properly structured Blockchain portfolio gives a person financial stability, financial growth, and the operational capacity to generate business income.

These are just some of the various Blockchain protocols that you can piece together to create your very own Blockchain Banking Operations. Take the time to do your own research into the various options that are available to you. You do not need each and every protocol listed, so be mindful of your own financial needs, and capabilities. One of the benefits of Blockchain technology is the ability to customized your financial tool box to fit your very own individual economic situation. Now that we are able to see how we can piece the Blockchain protocols together to form our own banking operations. Now, we need to take a look at the actual structure of this the proposed entity. The process of building the structure will be similar to the process of building out the financial operations. We will piece together the required technological building blocks to suit our needs as we see fit for operations.

No third-parties need, no central authority required, and no approvals sought after. Blockchain is all about owning your own data, owning your own property, and owning your own destiny. So building out a gateway to a blockchain platform will begin in the same place as a regular online business. The journey begins with the domain name. We will be using a domain name service that gives us the ability to purchase our domain name once, and own it for life. We need to own our domain name not just rent it. I will be using a web3 supplier called Unstoppable Domains. This will give us a domain name that we own, can use in coordination with a DAO, and receive blockchain payments directly to the domain name. Our web3 domain names gives additional functionalities to the website that we will build upon it. The website will be the visual gateway into the Rabbit Hole. It will function as a funnel into both the Blockchain investment, and the Blockchain social media worlds. This gateway will provide links to investments protocols, and a Rabbit Hole community built upon decentralized social media platforms. The Rabbit Hole gateway will employ the use of IPFS hosting using the Inter-Planetary File System Desktop node hosting application. On the back end of the gateway will be the DAO's that are connected to Blockchain Global Safes, and a list of Dapps (Decentralized Applications). The DAO's can be used to setup financial departments, communities of interest, investment vehicles, and community treasuries. Through the use of a DAO, and corresponding Dapps a service, or a community can become a NFT gated service, or a gated community for tiered services. This is a way to distinguish the value of the learning community from the value of the investment community. At this point you would just need to decide which particular DAO creation package you want to use, and what Dapps you can use to extend the functionality of your DAO, or DAO's. Once you have these structures in place, and ready to operate then you can focus on localization to turn your web3 site into a global gateway. Localization is making sure that your website, and content is translated into multiple languages, cultures, and contexts for the global markets, and visitors. Definitely, make sure that your content is customized for the global markets that you plan on operating within.

This is the point where deeper research into the various protocols that pique your financial interests. Do some background research on their history of audits, operations, and growth. Look at the community support of each particular protocol. Check out their long-term goals, and timeline. Make sure you understand the criteria of using each protocol before you choose to incorporate that protocol into your business plan. Start making a list of researched protocols for each banking services that you are interested in providing, or developing. Work through your process of elimination to narrow down the best prospects for your initial design. Feel free to try a few different model designs for your potential Blockchain bank. This will give you a greater understanding of the type of requirements, and commitments needed for various banking operations. Remember that your banking design can be changed as time passes so understanding various operational needs will be a great benefit to any future financial engineering.

CHAPTER VIII

Blockchain Banking Portfolio

Building A Financial Operations Portfolio With Blockchain

There are numerous methods, and strategies for building a Blockchain investment portfolio. The structure of your portfolio will depend on the answers to these following questions: *Why are you investing in Blockchain? What are your investment goals, and the objectives that you seek to accomplish? What are the levels of risk that you are willing to accept? How much money do you have to begin your investment portfolio?* Once you have the answers to those questions then you can decide which type of Blockchain portfolio is best for your needs. The type of crypto portfolio that you choose to build will ultimately have an impact on the capacity of your future business operations.

It may turnout that you feel more comfortable building a liquidity portfolio. This would be a portfolio that is compose of various crypto exchange tokens that you are providing liquidity to on their platforms. If your investment goals are to store your value, and hedge against inflation for a certain period of time. Then, it may be best that you have a staking-based Blockchain investment portfolio. This would be a portfolio of the best tokens to lock up in staking mechanism with the best terms for investment. You will probably have your token holdings spread out over several different Blockchains as the markets dictate.

For instance, to all my gamers that spend serious amounts of time playing video games should be building a GameFi portfolio. Because in Blockchain-based games the players actually get paid in tokens to play the games. They also get to own all of their in-game assets that they acquired through their game play. Blockchain gaming players can monetize both their time playing the games, and the gaming rewards as digital assets. This is called a Game/Finance economic model that is backed by tokenization. A GameFi portfolio will be comprised of various Blockchain game tokens of various gaming communities. Keep in mind that an investor can build a GameFi portfolio without actually playing the games. You will still have to be knowledgeable about the gaming ecosystem along with the primary, and secondary marketplaces associated with the gaming assets.

Some investors only invest in certain Blockchain ecosystems, and some investors only buy newly launched tokens. Some people have bandwagon portfolios. These are people who only buy "what's hot", and on the top of the markets. Yes, there are unicorn-only portfolios in Blockchain. It will depend on your investment goals, and your risk level.

For our Blockchain portfolio our investment goals are clear, we need to build a portfolio that will give us the capacity of **banking operations**.

This means our token holdings will be spread across multiple Blockchains, and various ecosystems. We will need to answer a second set of questions to start determining which tokens we need in our portfolio. We need to decide what type of banking operations that we want to establish. *Are we building retail banking operations, investment banking operations, or a combination of both? Are you building your banking operations around a certain group, or community like a credit union?* The answers to these questions will have a direct impact on your protocol decision-making process. The choice of protocols determines the design of your banking entity. The process of stacking protocols should reflect your initial financial goals, and objectives. This is how you will coordinate your original fiscal goals with the economic operations of a Blockchain bank.

Now it's time to get some clarification on the common types of banking operations. Most people are familiar with retail banking operations. A retail bank provides cash deposits & cash withdrawals, personal & business loans, money transfers, and credit services. These are the basic banking services that are provided to the masses. You can refer to these financial services as local banking operations that support local communities. Retail banks are the cornerstones of the local economies, and surrounding areas. These types of banks focus their operations in locations close to their place of operations. Retail banking operations help organize local resources, and community development.

Credit unions function in the same fashion as retail banking operations. The only difference is credit unions have narrowed their scope of operations to serve a particular sub-section of the local community. A credit union is basically a retail bank that has structured its financial operations to serve a specific demographic within the local community. The credit unions provides financial support to specific groups of people based on a specific business, church, occupational trade, location, or other limited demographics. Credit unions manage their limited scope of operations through a membership model. The only people who can access the banking entity are approved members of the credit union.

There is another type of banking entity that has a more globally expanded focus of operations. This type of bank focuses on customers that needs financial support to become competitive on a global scale. We call these types of banking entities, and their financial operations investment banks. Investment banks operate with a global focus on development, and management in multiple markets around the world. These types of institutions function as economic drivers of the global economies. Investment banks offer some of the same services as retail banks, but those services are scaled to meet the global needs of both the clients, and the global markets that they operate their businesses. For instance in investment banking, retail banking deposit services are scaled to become global treasury management services for corporations in the global market. Retail banks provide regular business loans, but investment banks provide more complex funding services for global businesses. Investment banks help companies raise funds via the stock markets, bond markets, and capital markets to address their funding needs. On a local level a business loan can help a company buy equipment to expand its business. Investment banking operations helps companies raise funds to acquire additional companies to add to their portfolio, and expand their business operations. They also provide advisory services, wealth management, and insurance services. Currency exchange, and Forex services are some other common services provide by investment banks. Providing currency exchange services helps facilitate payment settlements in the various global markets. These are just some of the services that investment banks provide clients in the global marketplace.

So decide what type of banking entity would be best for you to design according to your financial objectives, and capabilities. This is the point where your financial goals, your banking business plan, and your reality becomes a design for your Blockchain bank. We're almost ready to start choosing the actual pieces of the Blockchain Bank that we are building. And, we are about a half-a-dozen to a dozen steps away from completing our building process. But, before we can start researching the protocols that we may need for our bank, we need to have a strong criteria for selecting investment tokens in the various Blockchain ecosystems. We will be looking at several different types of tokens so we need a criteria that will help us determine a good investment from a risky endeavor. The criteria for investing in Blockchain has evolved over the past decade and a half. But, the basic steps remain the same.

The first step in Blockchain research is read the white paper of the token that you are interested in acquiring. The white paper is the technical write-up on the token with its purposes, and its stated functions. The white paper will tell you whether it's a entirely new token, or an improvement of an existing token. You will be able to see what types of Blockchains this token operates on, and the various ecosystems. Is it operating on popular Blockchains, or did they create an entirely new Blockchain to suit their technical needs? Does this token provide a new solution to the community, or is the token just crowding the field? What are the advantages of using this particular token utilities? Are there similar tokens with the same functions? What is the relationship between this token, and other tokens in the Blockchain ecosystem. One of the most important factors a white paper can tell you is the decentralization factors. Remember true Blockchain technology is decentralized in nature. We do not support a central authority using Blockchain technology as a mask of deception. The white paper should also have a breakdown of the token's tokenomics.

Understanding the tokenomics of digital assets, and cryptocurrencies is essential to building financial acumen. *Tokenomics is the combination of the total token supply, the issuance model (rate & method of distribution of the token), and the valuation of the token.* Studying these factors before investing will give you a strong idea of the overall health of each token investment. Each token that you invest in should have a fixed, and limited supply of tokens. When a token does not have a fixed limited amount of tokens then that token becomes a devalued digital fiat version...just like the inflation-based fiat reserve notes in legacy banking. When a token has a limited supply of tokens then the Law of Supply & Demand works in favor of all investors. Plus, you will have a financial timeline of asset maturity when you compare the total supply with the total amount of tokens in current circulation. The next aspect of tokenomics to inspect is the distribution model. What is the rate of issuance, and the velocity of the coin issuance? The allocation, and distribution of tokens in relation to the total supply will affect the token's valuation. So if you plan to launch a new token in relation to your new Blockchain bank you must invest deep thought into the tokenomics.

You will need to find a balance between the valuation/price, and the token supply to meet both the needs of the project, and the needs of the potential investors. A tokenomics design that has too many tokens dilutes the invested value, and not having enough tokens impairs both the needs of the project, and the financial mission. When the valuation of a token is too low this means the timeline for the investment to mature is too long to make it a valuable investment. An extra-high price valuation can make an investment unattractive, and price out a large majority of investors. This creates a problem for building community support. And, the vitality of the community surrounding the project can be indicative of the overall future health of the project. Keep these factors in mind as you move forward in the design process.

Now it's time to start deep diving into the protocols that you need to design the type of Blockchain bank that you have chosen to build. Whether your choice of a bank is local retail operations, global investment banking operations, or a custom tailored-made credit union banking operation. Revisit the previous chapter of protocol listings according to your desired banking design. Setup your research process, and schedule so you can be organized, and decisive in you choices. Researching these protocols for your bank will help you make well-informed choices, and grow your overall knowledge of the various

Blockchain ecosystems. This is how you will grow your valued assets, and your financial acumen. Imagine how many financial tools, services, and products that you will be familiar with once you have researched over 50 - 100 different protocols for your new banking entity. Depending on the scope of your operations this may be a low number of protocols to research, and discover. Determine how many protocols that you might need to build out each of the financial services departments. Look at how many different Blockchains that each protocol is operating on in the various ecosystems. Compare the prices, community support, and overall growth of each protocol against other potential protocols. Would holding this, or that protocol's token be a good investment outside of your normal banking operations? These are just some of the factors to consider during your research of protocols.

The researching of protocols can become a long process depending on the scope of your desired business operations. Plus, the amount of financial knowledge that you seek to acquire will factor into the amount of time that is invested into your protocol researching process. This is why it is important to organize all of the data that you collect in a way that helps to facilitate your ultimate goals. To accomplish this task I would suggest organizing your Blockchain protocol research into multiple lists, price ranges, and budgets. Organizing your protocol research with these simple methods will help you make the important operational decisions that will eventually become the basis of your operations. Think about the goal of your research, and what you will need to accomplish at the end of this process. Once your research is complete you should have a list of all the tokens that you'll need to start your banking operations. You're also going to need to know how much funding is needed to acquire these selected tokens. This means you need to know all the prices before you can create a budget. The token price ranges, and your budget will reveal your actual investment capabilities.

When you reach the point where you're clear on which tokens that you can initially afford then you can make the next important decision about your banking operations. You will need to decide on the order of operations. The order of operations will include decisions about which tokens to buy first, and which tokens to start putting to work in your operations. What tokens are you leading off with, and which tokens are best to supplement your initial investments? This is a process that you will repeat, and will refine on an ongoing process to manage your banking operations. This is how you will build your banking services, and departments. This process will help you build out a budget for each phase of operations from launch to the scaling of growth, and management.

The first list should be the longest list of all of the blockchain protocol lists. This is your feeder list of potential Blockchain protocols that you will need to break down into the needed information to make sound business decisions about your operations. This is the combined list of all of the protocols that you want to use, might want to use, or need to learn about. As you work through this list you will create an "already researched" list with protocol information such as utilities, prices, and banking service among other data points that you might need to make your decisions. The third, and fourth lists can be broken down by banking services & products researched, and by the token prices during the researched time frame. The next list will comprise all of the tokens that you have decided that you need for your banking operations. You will compare this list with your budget to form a list of tokens that you're actually able to acquire to start building your Block-chain portfolio. You will follow your initial purchasing list with your second phase, third phase, and fourth phase of purchasing lists. You can repeat this process as many times as you need to produce both the decision-making information, and the operational growth that you desire. This will give you an actual operating budget, and an order of operations. Now you have some operational data to work with such as price comparison data, funding needs data, and operational direction data. When you compare the data obtained during research with the data during purchasing phases you will be able to see the growth, and health status of each protocol in your operations. This assessment should be performed periodically according to your own specifications, or needs. There is only one more comparison that you will need to make before you start purchasing the needed tokens. Make sure to compare your original goals with your researched decisions to see if you have met all of the needs of your banking design goals. Once your design goals have been met then move forward to the token purchasing phase. If your design goals are not accomplished then make the necessary adjustments to either your process, or your design.

You are finally ready to start building out your Blockchain portfolio. The proper amount of sound research will give you a portfolio that will stand on its own even without actual banking operations. What I'm saying is your token holdings can grow without you actually putting them to work. For instance, there are a lot of tokens that can be staked to receive additional tokens. But, most people are comfortable profiting from just holding the tokens in their portfolio, and waiting on the price increases that comes with project growth kinda like company stocks. Good research can increase the value of your portfolio. But, knowledgeable research can put you on the path of generating wealth on a consistent basis. Please do your research, and choose wisely. Lastly, do not be afraid to incorporate tokens into your portfolio that are not banking operations related if you feel they can help you achieve your ultimate goals. Remember that you have options for short-term, mid-term, and long-term investment goals. This is why we order our operations so we can know what moves we're making first, and what the next steps will be in the future. Once you complete this necessary research then it's time for you to pull the trigger, and start getting you some tokens for your banking operations. Good luck my friend!!!

CHAPTER IX

Customization & Extending Banking Capibilities

Customizing & Growing The Functionality Of Our Web3 Platform

In this chapter we will be focusing on building out the actual structure that will deliver the various banking services. We will be diving deep into the power of DAO's, Dapps, and Blockchain social media platforms. Then we're going to look at ways to format, and prepare our web3 portal for the global stage. You cannot build a global banking platform in only one language, or one country. This means we need to have a method of localization to cater to the global markets. All of our content needs to be translated into the native languages, and cultures of our website visitors, and the markets where they reside. We will use DAO's for the structure of our Blockchain banking platform. We will also incorporate various Dapps to extend the functionalities, and capabilities of these DAO's. Then we are going to use Blockchain-based social media platforms that will build, and empower the community that we build around our banking platform. This will be the means that we will use to setup our delivery system.

Lets start with the **Decentralized Autonomous Organization (DAO)**, and how it differs from a regular organization. Most organizations that people are familiar with, and join are built on a top-down centralized authority structure. This means all policy, decisions, and authority are facilitated from the individual who occupies the head position of the organization. Legacy organizations are based upon a pyramid structure with a few powerful people at the top who rule over the rest of the members in the organization. Decentralized Autonomous Organizations are structured, and they function in a contrary fashion to a pyramid ruling structure. DAO's operate with a horizontal structure that includes the majority into the governance process. This type of organization keeps the entire decision-making process decentralized from governance to implementation. This process insures that the organization can remain autonomous without individual leadership as a headless organization. You can format an organization that includes all of the members interests regardless when they have joined the organization, and include them directly into the governance process. Using DAO's to build our platform reflects the decentralized nature of the communities that embrace Blockchain technology. It's an improved method of organizing people, resources, and businesses.

Creating a DAO may seem like a confusing, and complex process. But, the truth is you can start a DAO easier than starting a conversation about starting a new organization. The Blockchain community has made great strides in the area of DAO development. There is no need for you to know Blockchain programming to create, and manage a DAO nowadays. You can choose from a handful of DAO frameworks that will provide you with an easy startup package. Think of it as a quick starter kit for creating a DAO. This creation process starts the same way as every other Blockchain process, with some in-depth research. The first decision that you will need to make is which Blockchain will be best for your DAO. Both the Blockchain, and the DAO framework that you choose will have different pros and cons for your project. DAO frameworks will offer different functionalities, and operations. The same holds true for the Blockchains that you choose for your DAO.

Here are some factors to consider in your decision-making process before creating your DAO. Compare the cost, and speed of transactions on each of the Blockchains. Look at the scalability factor for both the Blockchain, and your own crypto project. Because the scalability factor will affect the future growth of your project once it's up and running. Then, you will have to look at that one factor that affects all of us...security. After you look at the potential Blockchains for DAO creation then you need to compare the DAO frameworks for the potential benefits to your projects.

Here are some of the most popular DAO frameworks to research: Gnosis, DAOstack, Aragon, Colony, and DAOhaus. Every DAO will have its own needs for the project, and its treasury management responsibility. Some frameworks are better for managing DAO treasuries by using multi-signature smart wallets. The use of smart wallets can help streamline a DAO's management of assets, and financial distributions to members. Allocation of the organization's resources to its members might be a priority for your project on a regular basis. You may decide to use a DAO framework that provides additional options for collective governance. This might assist your organization with the long-term planning for the continuity of its operations. You may not need a lot of advance features for your DAO, and choose a more simplistic framework model with just the basic features. Other frameworks will allow you to structure your DAO with sub-DAO's, and sub-treasuries. Do the necessary research to discover all of the options that are available to you when customizing the functionality of your DAO.

Each of these frameworks has a different approach to treasury management, governance, and additional customization options. These customization options come in the form of **Decentralized Applications (Dapps)**. You can think of Dapps as open-sourced applications that are built upon Blockchain technology. There are numerous types of Dapps available to DAO developers. A DAO can be created for the purpose of managing an organization, or to start a new business. So the Dapps that you choose will help facilitate the process of achieving your ultimate goals. In a DAO, the governance mechanism is one of the most important aspect of your project. You will have an abundance of Dapps that will assist you with the voting of members. Some Dapps will manage the actual voting process while others will help with the management of proposal submission for group voting.

There are Dapps that deal with creating legal documents for your DAO. You can customize your DAO to perform a series of financial transactions including fundraising, allocations, and payments. You might need a way to convert fiat money to a digital assets, or off-ramping crypto back into fiat notes. Some of the most important Dapps are the applications that facilitate collaboration amongst members and the DAO. This could involve both investing, or splitting the proceeds of a recently sold NFT. Some of the needs of your DAO can be automated to ease the burden of managing multiple processes. There are hundreds of Dapp options that are available to anyone looking to customize a DAO. Try looking up a list of DAO tooling options on alchemy.com, or dappradar.com. A basic search of Dapps, and DAO tooling will yield even more options for customizing your DAO.

After you know how to setup your DAO's structure, and how to add various customization features then you can setup a structure to meet your project's needs. The next step in this new banking design process is to build a supporting community around your Blockchain project. So now it's time to build a solid social media presence for your new Blockchain project. Our goal isn't building an outdated legacy project with legacy technology. We are building a new type of web3 platform with the benefits of Blockchain technology. Therefore it only makes sense to be building our new community on social media platforms that utilizes web3, and Blockchain technology. Fortunately, there are numerous options available to us that offers the type of benefits that legacy social media companies will never provide people.

Almost everyone is on some kind of social media platform, but none of these people have any meaningful control of the data created on the platforms that they are using everyday. Outdated social media platforms are built upon a mass exploitation model to serve the private interests of a few companies to maximize their Wall St. profit margins. In addition to, stealing all of the value from the content creators; these private companies also exercising an unconstitutional level of control over the people's freedom of speech, and personal expression. The creation of content is also a process of creating digital assets, and generating value. Legacy social media platforms forces social media users to produce content while the company produces the profits. How many times have you seen a social media user have their hard work, and content confiscated via being banned from a privately-owned platform. The private social platform stills has access to this content, but the content creator loses complete control of their digital assets. This legacy model of building social communities is unsustainable, and unfit for the future of our new web3 project.

Do you know what the following names have in common: Mark Zuckerberg, Zhang Yiming, Daniel Ek, Jan Koum, Brian Acton, Evan Spiegel, Leonid Radvinsky, Pavel Durov, and the next-Enron-dude who is destroying the platform formerly known as twitter? All of them are owners of the top social media companies that exploit user's content for personal profit. Neither one of them has had a content creator become a billionaire off of creating content on their platforms. So that shows you how much content value is being stolen from each and every user on their platforms. It also highlights how much user control, and ownership is being forfeited by traditional social media users.

There are some other commonalities present amongst these private platform owners. Each of them has a fear that the people of the world will realize one day that they do not need any of there private social platforms to build strong social communities. They fear social media users will wake up, and start using decentralized social media platforms to start asserting their rights to acquire sustainable wealth on their own accord. These owners know their plantation models of social networking can crumble once social media users decide to take back control, and monetize their content creation process without third-party interference. It's only a matter of time before there is a mass migration from closed-social media platforms to community-owned social media platforms. Once enough people discover there are great alternatives to Twitter, Facebook, Tik Tok, Instagram, VK, Snapchat, Linkedin, and other outdated platforms then power and wealth will be shifted back to the community members.

Well we're not going to wait for the mass exodus movement away from oppressive social media platforms. We will be using the newest forms of technology for building viable social communities using Blockchain-based social media platforms. The type of social networking platforms we will be using will give the individuals the ability to protect their privacy, and data from being sold to other corporations. The community users will not have to give up all of their personal identifiable information to register with the social networks. And, the user's activity is not tracked by a predatory centralized authority. A decentralized social platform guarantees a censorship-free platform that guards your freedom from both corporate, and governmental interference. This is possible because the data is saved on a decentralized network of servers instead of a centralized storage system. All of the created content can, and will be owned by the creators instead of the platforms.

These decentralized platforms are built upon Blockchain technology, and gives the content creators the ability to monetize their social networking activity. When you own your own data then you should be able to profit from the value of that social interaction. This means everyone should have the option to get paid for meaningful posts, and social interactions that bring value to the social network. Blockchain technology, and the power of tokenization enables social media posts to monetized via NFT's, or social reward tokens. Blockchain-based social media platforms allow community members to reward other members with redeemable reward tokens for their valuable contributions to the community network. This type of social media interaction encourages meaningful contributions instead of the traditional method of spamming social networks to gain enough followers to meet the monetization quotas of private social media companies. A lot of people are familiar with Patreon's platform, but Patreon is just the tip of the iceberg.

The social media revolution runs extremely deep within the Blockchain communities. Blockchain developers have been busy creating viable alternatives to every major social media platform the exists in the market. There are multiple options for Twitter, Facebook, Youtube, Instagram, Linkedin, Tik Tok, and other popular platforms. Not only are there individual options available to social media users, but there is a group of Blockchain developer that are dedicated to creating a social networking revolution for society. These Blockchain developers went far beyond creating an alternative to popular social media platforms. They actually created an entire new Blockchain dedicated to decentralizing social media. They called it **DeSo** which stands for Decentralized Social. They created a Blockchain ecosystem that empowers developers to create decentralized applications to decentralize social interactions across multiple networks without the need to keep creating a new profile for each network. You created one profile that you can use in all of the Twitter, Facebook, and Linkedin alternatives. And, if you don't see an alternative to your favorite social media platform then you have the option to be the first one to build that Dapp for the entire community.

Let's take a deep dive into the world of DeSo, and the social platform alternatives that are available to the community for web3 social interactions. We'll start with some of the most popular alternatives being used on the DeSo Blockchain. Then as usual, you can do the required research into all of the various web3 options available to social media users whether on the DeSo Blockchain, or other individual alternatives. Take a look at the following DeSo web3 decentralized social media platform alternatives:

- **Diamond-** a decentralized web3 Twitter alternative where you can earn for your engagement
- **Videso-** a decentralized Youtube alternative for posting, and watching videos
- **DeSo Chat Protocol-** a Blockchain encrypted messaging alternative to Telegram for direct messaging & group chats that you can attach a crypto wallet
- **Zirkels-** a decentralized Medium alternative that is ad free, and pays you money
- **Stori-** a web3 alternative to Tik Tok that helps both creators, and fans earn for their engagement
- **Pearl-** an Instagram alternative for creators to earn with their fans
- **Entre-** a decentralized web3 Linkedin alternative for business networking
- **Vibehut-** a decentralized Zoom alternative to make video calls, and conferencing
- **Mousai-** a web3 alternative to Spotify for musicians, and artists to earn a living
- **Pineso-** a decentralized alternative to Pinterest for posting, and exploring images on DeSo
- **NFTz-** a decentralized alternative to OpenSea for NFT marketplaces
- **DeSocialWorld-** a decentralized social network that enables you to post in multiple languages just by clicking another language on the side menu

The last option on this bullet list (DeSocialWorld) is a great localization option to reach beyond your local communities, and markets. One the of the greatest failures of most businesses is single-mindedness, and narcissism. In business it is not about you, and your personal language. It will always be about the customers, their cultures, and the languages that they speak. We live in a globally-connected world that is full of beautiful, and diverse languages that are spoken in numerous markets around the world. You cannot prepare your business for these global markets without preparing your business for the languages associated with these markets. So we need to figure out the best way to go about achieving this goal.

We need to setup a localization process for our web3 portal, and all of our content that will be created for the global markets. Localization is basically a language, and cultural translation process for all of our content whether it's a website, or marketing material. For instance, when you go to a website and you have the option to choose your native language this means the website has created a localization process for both their content, and you the customer. Both Blockchain, and banking are global technologies that are consumed by a global audience. We need to play to the audience's sensibilities to the best of our abilities. So lets check out some of the localization options that are available to customize our Blockchain banking project.

We need to decide what will be best for us both initially, and on an ongoing basis. There are translation companies that offer their paid services on the market. There are software options that might be a great fit for our business ventures. Some of the software is open-sourced, and some software require subscriptions, or licensing to be used for business. The budget that is available to you will help determine the best options that are available throughout each phase of your project. The size, and scope of your operations will be another determining factor. A larger operation will definitely need a more robust localization process. If you have a substantial budget then a simple online search for the best translation services may be sufficient to meet your organization's needs. But, if you are bootstrapping your startup then you might need to look into both paid, and open-source translation tools that are currently available on the market. It's going to take some time, and research before you can implement your localization process. Here are some of the factors that you need to keep in mind before I list some localization options.

There are various computer, or machine translation tools in the translation, and language services industry. There are some tools that only focus on terminology translation. Some tools have a degree of automation incorporated into their translation capabilities. Then there are TMS (Translation Management Systems) tooling options. Some of these tools will only streamline the process of translation, and will need some followup by others to complete the localization process. Other tools will assist with the confirmation of the translated information. Some tools have memory retention capabilities so you only need to translate the same content once. And, then there are those tools that only help translators do their job a little easier. But, nothing can compare to actually having a person who is fluent in both the language, and culture for your translated content. You might start with some tools of your choice, and then hire someone off Fiverr to confirm, and tweak your translations. The proper amount of research will help you create the best localization process for your project.

Here are some starting points for your research:

- **Okapi**- an open-source localization framework of various tools to assist in translating your content
- **Tolgee**- open-source tool with various capabilities
- **Glossia**- an open-source localization operating system
- **Mojito**- an automation platform that enables localization
- **Pontoon**- helps to localize Mozilla search engine
- **Weblate**- another translation tool

I would suggest you start searching terms such as: localization, translation management systems, machine translation, localization automation tools, open-source translation tools, translation services, translation plugins, localization process, etc. Look for the best solution that will help you customize your project for each foreign market. Remember you do not have to translate all of your desired languages at once. Both building your business, and adapting to markets is an ongoing process. Hopefully, your research will yield a lot of additional options to customize your future, and current projects. Good luck my friend!!!

CHAPTER X

Locations, Launching, Let's Go!!!

Blockchain is an global technology that is being adopted in virtually every country on the planet. And, we know there are numerous views about politics in each of these countries concerning their own collective interests. The same can be said about the global views, and collective interests of Blockchain technology. When I speak of the global adoption of Blockchain, I speak in the terms of the general population adopting Blockchain technology not their governments. When you compare the top countries for crypto adoption, and the top crypto friendly governments you will have two very distinctive lists. One list will show you the will of the people while the other list will highlight governmental self-interests. Some countries are the enemies of Blockchain technology, and the benefits that this technology provides the people. You have some countries that profess to be crypto-friendly, but by looking at their laws you can't accept their proclamations as truth. Then you have countries that want to be the leaders in the Blockchain industry, but are totally clueless as to what is needed to properly accomplish this goal.

Most countries only know how to tax, confiscate, and control the wealth of their citizens. They function as large mafia-like organizations that demand tribute to be paid of anyone making a living in their controlled territories. Because Blockchain technology is creating a global paradigm shift there seems to be an arms race to over-legislate the future of Blockchain before they even understand what is happening in the present. It only reminds us that governments do not have the interest of the people at hand. Governments are in the mismanagement business. Extortion for self-survival is their business model. The best, and ultimate solution may be to form our own Blockchain-Nations!!! The people just need to realize that all of the needed technologies already exist in our present day societies.

The failed attempts to over-regulate a new paradigm-shifting technology is not new to society. We've seen these same efforts recently with the adoption of the internet. People forget that the early internet adopters operated in an online tax haven before failed states polluted, and corrupted the beauty of that paradigm shift. This is how amazon.com was able to amass its early fortunes during their startup stages. There was a brief period of time before central authority was allowed to destroy the Internet's decentralized movement. Blockchain is more than just a freedom technology. It is a decentralization revolution that has little to no room for centralized authority. So where to start your Blockchain project may be a mute point, but we will discuss it because people's conditioned mentalities has yet to evolve to the point to realize freedom is a choice. This is why Harriet Tubman is one of the most famous people on the planet because she spent her life highlighting how slaves will cling to a system of slavery even when they are given a choice to be free. This is because they cannot imagine living in a better world where they take responsibility for their own actions. Blockchain technology provides you with the tools to take these necessary steps forward, and upward.

So lets take a look at the good, and bad options that are available to the early adopters of Blockchain technology. Hopefully by comparing both types of regulation you will be able to have a greater understanding of the landscapes. The subject of "where is the best place to start a new Blockchain project" is constantly changing, evolving, and devolving. Countries that were one of the best places to start your Blockchain project have become the ones chasing hundreds of crypto businesses away by enacting oppressive crypto laws such as Singapore. Take a look at Singapore's crypto-legislation in 2023 versus the legislation that was enacted in mid-2024. Over hundred applications for crypto permits, and licenses were withdrawn by Blockchain businesses due to the enactment of their new regulations. This is an example of a countries that was originally crypto-friendly, but then the tiger changed its stripes.

Lets look at an example of a country that professes the goal of becoming the most crypto-friendly nation, but has no idea, or experience to accomplish this goal. The Marshall Islands is an example of another country that had great potential, but totally dropped the ball completely. Most people don't realize that the Marshall Islands is a nation where most of the global shipping companies have their businesses registered due to their favorable legislation for the shipping industry. Evidently, the islands had experienced industry operators in the shipping industry that helped craft their laws to benefit the shipping industry. But, it is more than obvious that the Marshall Islands did not take that same approach when it came to the Blockchain industry. The same respect was not accorded to the crypto community.

The reason that the Marshall Islands popped up on the crypto radar was an announcement that the nation was going to craft legislation for decentralized autonomous organizations (DAO's) to have an official legal recognition as a business entity in their country. Basically, DAO's could be registered just like a LLC entity with full legal protections in their nation. But, when it was time to develop that legislation process they decided that their personal self-interest was far more important than the interests of the crypto communities that they professed to serve. All of their legislative process was decided by one legislator, and one self-serving outsider. A person who went from being an anal cancer volunteer came to the islands, and became their crypto-jesus.

One outsider convinced a local legislator to create their own registration DAO, and force all future DAO registrations to go through them to establish a "legally recognized DAO" in the Marshall Islands. These two individuals created laws that said the registration fees will be on a sliding scale depending on how valuable your Blockchain project is considered. This sliding fee scale wasn't just for the initial registration, but on an annual basis according to the growth of your Blockchain business. Somehow, these two characters felt they could enact laws that basically turn MIDAO into a venture capitalist business without supplying capital. Their registration process would essentially make MIDAO part owner of every DAO that is registered in the Marshall Islands. Just think about the audacity of self-serving greed that it takes to write a law that combines both mafia, and IRS practices. Then, the Marshall Islands proposed to start their own crypto-currency as legal tender along side of the dollar. So, eventually all of these ongoing DAO registration fees would have to be paid in their new token costing registration TO increase. No nation deserves an ongoing percentage of your business's success in exchange for poorly written words. What this nation professed it wanted to become, and what it actually became are to separate realities.

Regulation should provide a healthy space to operate freely, and safely. It shouldn't mean having a monopoly on extortion, and taxation. Nations better be extremely careful before they play themselves off the Block. The Blockchain community has already proven that we do not need any outside regulators because we are a decentralized solution-based community driven by collective results. Beware of legacy frameworks that cannot be super-imposed upon new technology. Remember when the first cars were invented, people tried to operate them on roads that had been degraded by horse-drawn buggies for over a century. It was a bumpy ride that appeared to look like a failed technology. The only failure was attempting to introduce a new technology on top of a legacy framework. There were a lot of accidents, and injuries because the roads were not built for the new technology. The governments tried to convince the people the new technology was dangerous, and needed regulations because the foundations were built for an outdated legacy transportation framework.

You cannot cast fertile seeds upon freshly laid concrete no matter how moist the ground looks. The truth is Blockchain technology is a global boycott movement against failed oppressive outdated centralized authority schemes. We need to fight the Good Fight through every phase of this movement. Below is a list of locations to research, in regards to, potential places to startup your new Blockchain business. They're not listed in order of preference due to the constantly changing 'legal' landscapes.

1. United Arab Emirates/Dubai
2. Malta
3. British Virgin Islands
4. Seychelles
5. African Countries
6. Estonia
7. Bermuda
8. Luxembourg
9. Portugal
10. El Salvador

Dubai is a leader in virtually every field of commerce on the planet. This also includes the Blockchain fintech industry. Dubai is a crypto-friendly location that has a very robust financial center that is capable to compete on the global financial stages. Dubai is also known as a technology innovation hub, and tends to embrace all forms of new technologies. The UAE is also a great location for fiat to crypto conversion for on-ramping, and off-ramping investment funds. In fact, Dubai has some banks that allow customers to buy crypto assets directly from their personal bank accounts. Dubai is also one of the premiere investment locations on the planet. So there is an abundance of surplus capital looking to be invested into the next big thing. Plus, its location provides easy access to Asia, Africa, and Europe. Dubai can be a great location to start a Blockchain Bank!!!

Malta is a location that some people might find surprising to see on this list of favorable startup locations. But, Malta was one of the first nations to embrace the crypto community. Malta started working on its legal frameworks for the crypto community back in 2018. Malta was a pioneer in providing a safe environment for Blockchain companies to operate. This is how Malta became the home to one of the largest crypto exchanges on the planet. A Blockchain unicorn company called Binance established its home on this little Blockchain island. In addition to embracing Blockchain, Malta has a history of providing a safe space for financial freedoms for global businesses.

British Virgin Islands is a good location to consider for starting a Blockchain business. BVI has a strong legal framework that protects all businesses that operate within its territories. Plus, the British Virgin Islands is a crypto-friendly nation on top of being a pro-business friendly nation. BVI has a well established culture of protecting businesses from governmental over-reach, and oppression. BVI is known for protecting the privacy of both business owners, and global investors. This is an added benefit on top of being a crypto-friendly nation. Research the benefits of starting your Blockchain bank in the BVI.

Seychelles is another Blockchain island off the coast of Africa that can be a great place to start your Blockchain banking business. Seychelles is kinda like the African version of the BVI. Seychelles has been a top business hub on the planet for a long time. This island is known for its privacy protections for business owners, and investors. The island has a strong business, and legal framework that welcomes businesses from around the world. Seychelles is a extremely crypto-friendly nation that is home to Blockchain unicorn businesses. This location offers a multitude of business benefits including a simple streamlined process for company formation. Do not overlook this beautiful African nation as a great location to possibly startup a crypto business.

African Countries are usually unjustifiably left off the list of best places to start a crypto business. The continent of Africa houses the biggest markets for multiple Blockchain use-cases. It has the most fertile ground for technological innovations across the spectrum. You will find more technology hub centers in Africa than any other locations on the planet. In fact, Nigeria probably has the largest, and oldest crypto adoption community of any other nation. African has the greatest needs for innovation so its embrace of paradigm shifting technologies is extremely steadfast in its commitments. Crypto was used to hedge hyper-inflation in Africa long before other nations started suffering from crippling inflation. Africa offers a lot of promising locations for future Blockchain businesses. Africa even has a country that uses crypto as its legal currency. Plus, venture capital's focus on Blockchain, and Africa is outpacing Blockchain investments in the United States. You better do your research before you get left behind.

Estonia is a top Blockchain country for development, innovation, and implementation. This is one nation that didn't hesitate to incorporate Blockchain technology into its infrastructure. Even their voting systems, and their citizens benefit from the technology. Estonia is a crypto-friendly nation that has a strong Fintech development community. Estonia is also known for their business friendly E-residency program. This program allows foreign citizens to apply for E-residency, and once approved you can start a new business in a few simple steps by applying online. This platform is a great way to gain access to the Euro-markets. I used this process personally to start my own consulting business in Estonia. So make sure Estonia is on your research list.

Bermuda is another nation striving to become a Blockchain-friendly island. They have been working hard to welcome Blockchain businesses to their beautiful island. They have established digital assets laws that recognize the crypto community's needs for protection. Bermuda is seeking to diversify the islands economy from just the tourism industry. And, their efforts to achieve this goal has put the beautiful island on the runway to takeoff as a crypto haven. Bermuda actually introduced legislation that created new standards for their banks to provide for Blockchain businesses. Bermuda has always been a business friendly nation for the international business community. Now it wants to do the same for the Blockchain businesses.

Luxembourg is widely known as the major financial center for the Euro zone. Some of the world's top financial services are run out of this nation. Luxembourg has been a worldwide leader in the financial services industry for years. Now, Luxembourg seeks to maintain that leading position by adding a framework that is welcoming to the Blockchain business community. This nation is striving to become a leading location for the new financial community that is coming of age. Luxembourg may become a major crypto hub as well as a major financial hub in the near future.

Portugal is a nation that usually gets overlooked as a crypto-friendly nation. Yes, it is known to some business leaders as a great place to incorporate a business because of its tax-friendly business laws. But, Portugal is not just a business friendly location it's also a crypto-friendly nation. One of the top Blockchain ATM manufacturers in the industry calls Portugal its home. This is another nation that would like to become a major hub for Blockchain innovation. The business benefits of starting a company in Portugal will also apply to new Blockchain startup companies. Research their laws to see if incorporating in Western Europe may be a good choice for you.

El Salvador is definitely a country that is crypto-friendly. They passed laws that made Bitcoin their legal tender for the entire nation. They even gave their citizens some Bitcoin in airdrops. El Salvador is one a few nations that uses crypto as their national currency. This is a movement that comes directly from the current president of their nation on down to the citizens. This nation is hedging their entire future on the benefits of Blockchain technology. El Salvador may become the top crypto haven in the western hemisphere. Make sure you add El Salvador to your research list.

When researching locations look for transparent frameworks that are easy to comprehend in basic terms. You are looking for places that only want to protect these new classes of digital assets not confiscate your assets via legislation. They should be extremely clear about what business registrations, licensing, or permits are needed to operate in their nation. Stay away from regulations that seek to demonize certain types of crypto products, or services. Any laws that are based on taxation of technology does not have the Blockchain community's best interest at heart. You are looking for locations that already have a history, and culture of providing financial freedom. Look for nations that will protect the privacy, the businesses, and the investors of Blockchain technology. These locations need a strong legal framework to backup their words, and desires. Remember it's not enough to have a great idea, and a strong team if you cannot protect all of the hard work, and success that you'll have achieved over time. Location is extremely important because having a safe protected home is just as important as building a home for a beautiful family.

Once you find a location that meets all of your criteria then reach out to the various financial bodies in each nation. Find out what is the actual process of getting up, and running in each nation. Inquire about the application process. See if financial licenses are needed. You will need to know if you need to travel to that location, or if they have an online process for registration. Making a list of questions, and a phone call can end up changing your life forever. Plus, it shows that you're a business that is serious about doing your due diligence. And, it could save you money, time, and sacrifice. Let's go, let's get this list going so you can launch!!!

CHAPTER XI

Sourcing Funding For Your Bank

Understanding Various Funding Options

There are a few elements that are essential to start, and succeed in business ventures. Those necessary elements are passion, commitment, and funding. Passion is great, but being passionate consumes huge amounts time. And, commitment is the measurement of dedicating time in a focused direction. For humans time is the single most valuable element that we possess in our lives. Everyday we spend more time than we had yesterday. This is the natural origin of the Law of Supply & Demand. We are constantly reminded of how valuable our time is in the world. So when we dedicate our focus, and time to a particular task we are conscious that we are taking time away from other aspects of our life. This is how we came to the conclusion that time is money. This is especially true when it comes to starting a business because a business is not a temporary process. Business is an act of continuity, or to be more specific it's a constant venture in time consumption. Therefore, one can conclude that funding will be needed through every phase of building a business. Business changes lives, and all life changes costs money whether those changes are positive, or negative changes.

The success, and failure of a business will be decided for all businesses by the amount of dedicated funding regardless of the type of venture. Almost every successful business story that you aspire to emulate needed some funding to start, or grow at some point. This funding aspect includes the amount of funding, access to funding, and the timing of that funding. Even the kind of funding that is needed for the business will be a determining factor in its success. Well we're building a global bank my friends so we will need funding worthy of a global venture. If you have decided to build a local bank then you will need funding that supports your needs to service your local communities. Most people fail because they look at funding as a one-time money shot that will hit, or missed. You should be thinking about funding through every phase of growth no matter how small, or big that next jump will be for your business.

Before you plant that first seed to start your business you might need pre-seed funding to organize,and gather resources. When it's time to actually start your business you will definitely need some adequate seed funding. Then, once you get your business up, and running you will need some growth & expansion funding to take your business to the next level. After, you have your business booming you'd better be thinking about future funding. Because there will be new employees, new markets, new equipment, and new ways of doing business that has to be funded. For all of the people who missed it, whenever you start a business you are actually starting two businesses. The business that you are passionate about, and the business of funding your passions.

The same way that you broke down your business concept into a business plan is the same way you should be breaking down your current, and future funding needs. In the startup world we describe these phases as pre-seed funding, seed funding, series A - C funding, and then series D representing future cash flow funding. Pre-seed funding covers all of the costs of needed to pull all of the pieces together. This is the costs associated with the pre-game work where you're putting the playbook together, assembling the team, and pricing the uniforms & equipment needed for the opening game. The seed funding is actually acquiring all of the chess pieces needed to play, and win the game. You can consider series A - C funding as the middle-game. This is the second, third, and fourth quarters of the game where you'll be switching out personnel, adapting to injuries, and demonstrating intestinal fortitude. Series D funding is for the post-game interviews, and preparing for the next levels. This is for the big decisions when the playoffs, and the championship game are in sight.

At this level the funding needs are completely different than the needs of the early funding stages. So when you are planning your business you also need to be planning your funding pitches for each level of your business planning. In the startup industry this is called a pitch deck. The phases of funding will vary depending on your business needs. Some funding will be required for product development, some funding will be needed for the expansion of the business, or for the entry into new profitable markets for your business. Understanding your funding needs will position your business for a successful journey.

There are several ways of funding your new business venture, but it is important to understand how each funding method will affect you, and your business in the future. Because we are building a Blockchain project I will discuss the multiple ways that you can raise funding with Blockchain mechanisms. Then, we will take a look into the venture capital industry, and pitch events. We will also look at some more familiar methods of funding such as grants, crowdfunding, and loans. Once you compare the pros, and cons of these funding methods then you can make the decision which is best for you.

The question to keep in mind is, "What are you willing to sacrifice for your funding needs?" Because, this will affect the type of funding you should pursue. Understand that giving up equity, and shares in your company for capital means giving up ownership in your company to strangers. Meanwhile, applying for business loans could mean giving up your personal timeline for the lenders repayment timeline. All lenders will require you to repay the loan according to their required timelines instead of the business's organic timeline. Make sure you stay true to yourself, and don't trade your values for valued capital. Think outside of the box, and try to fund as much as possible before you ask others for their money. When you think of funding think about whether you are selling off pieces of your company, giving up the control of the business, or just funding your business. Because all funding isn't friendly in fact some funding is straight up predatory.

Personally, and financially I love the power of Blockchain, and the flexibility of the solutions that this technology provides its users. There are multiple methods for funding Blockchain projects using various Blockchain mechanisms. And, new ways are constantly being developed to address issues with the previous methods, or needs. We will take a look at some of the ways Blockchain projects have been funded in the various Blockchain ecosystems. Some of these funding methods have produced unicorn companies with a $billion plus valuation. Some of the most successful Blockchain funding methods in the past few years involve some form of tokenomics. These included an Initial Coin Offerings (ICO's), an Initial DEX Offerings (IDO's), an Initial Exchange Offerings (IEO's), a Security Token Offerings (STO's), an Initial Game Offerings (IGO), and an Initial Liquidity Offerings (ILO's). All of these listed methods of fundraising have been used to build a new service, a product, or an innovation. They may look similar, but they function in totally different ways. Each of these methods have their own pros, and cons for each project. So make sure you understand the various mechanics of each funding process.

The initial coin offering is probably the easiest method to understand, and explain to the average person. An ICO basically is the Blockchain version of an Initial Public Offering (IPO). Both the ICO, and IPO fundraising mechanisms are early stage public capital raising efforts for business projects that rewards their early supporters. An Wall St. IPO raises capital by offering company stock shares for sale on the public stock market. These new stock shares can be purchased for a lower price in the earliest stages of the public offering, and then re-sold, or held for enormous profits once the company becomes valuable. An ICO can be considered the tokenized version of an IPO. Instead of offering stocks to the public an ICO will mint a token native to the specific project that will facilitate the raising of capital. Early supporters of the project can obtain the projects coins for a lower token price, and then as the project matures the value of those initial digital coins increases on public exchanges.

The ICO became famous a decade ago with the Ethereum ICO launch. Ethereum's 2014 ICO launch lasted a total of six weeks, and raised between \$18 - \$20 million for the project. The initial price of their Ethereum token was just over \$0.30 cents for one Ethereum token. During this six week time period one Bitcoin could net you between 1300 - 2000 tokens of Ethereum. Today one Ethereum token is valued at a couple thousand dollars. This made a lot of early investors wealthy, and Ethereum, and ICO's became extremely popular. There was another ICO funded project that generated a few billion dollars during their launch. The successes of ICO launches created a period in Blockchain history known as the ICO bubble. Everyone started launching ICO's even people who didn't have Blockchain projects to backup the fundraising process. This highlighted a greater need for investor protections, and safeguards for the crypto community as a whole. This started the evolution of funding phases to created a better tokenized method of fundraising.

The early risks of token-based fundraising created a need for more security in the fundraising process to defend against exit scams. This is how the Security Token Offering (STO) came into existence. Investors wanted some guard rails to anchor their initial investments. STO's became the next crypto-funding method that allowed projects to raise funding in exchange for equity in the Blockchain business. Offering equity through the issuance of security tokens provided a mechanism to offer ownership to initial investors. The transition period between ICO's, and STO's is when government agencies started campaigning for Blockchain mechanisms to be regulated like other financial instruments. This made STO's the most compliant, or the most over-regulated aspect in the decentralized community. This did not adhere to the decentralized nature of the Blockchain community, and the majority of crypto adopters began to look for the next solution to fund their projects without the negative affects of political intrusion.

It didn't take the crypto community too long to create a better solution for the funding of Blockchain projects. The next solution would evolve along side of the evolution of the tokenomic's basic infrastructure, the exchanges. The crypto exchanges are the crossroads that connect the economic flow of tokenomics. By 2019, crypto exchanges were the most trusted platforms in the Blockchain community. This is how Initial Exchange Offerings (IEO's) became the champion of the token-based fundraising methodologies. Almost everyone in the crypto community agreed that IEO's was the best solution for fundraising due to the benefits that it offered. The growth of the crypto exchange's user base demonstrated the trust, and security that the community had in this type of infrastructure. So during 2019 the idea was put forth to use the crypto exchanges as a conduit to offer the new coin mechanism. They were trusted, and they were more focused on security than any other entity in the Blockchain ecosystem. Plus, the exchanges could tap into their vast customer base without any extra effort. An IEO was considered the perfect marriage between Blockchain tokenomics, and infrastructure. In fact, in 2019 it was estimated that 95% - 98% of all token offerings were IEO's. There was only one problem that the crypto community had with IEO's, and that was making sure the exchanges did not become the central authority in the community.

Every year brings about new revelations, and new solutions for the crypto community. This is a natural occurrence with any new technology is constantly evolving to suit the needs of the people. In 2020, there was a major civil war raging inside of the Blockchain ecosystems. There was a rise in the number of centralized businesses competing for the Blockchain user base. A number of privately controlled crypto exchanges were challenging the original decentralized Blockchain models. Outdated legacy business models were trying to graph onto the Blockchain business momentum. These centralized exchanges began to function as conduits for different governmental regulatory bodies. So staying true to the Blockchain Gospel, the Blockchain community decided that a new solution was needed to protect the decentralized nature of our movement.

This came in the form of Initial DEX Offerings (IDO's). The community made a hard pivot from centralized exchanges to Decentralized Exchanges (DEX's). This community decision gained major momentum in correlation with the emerging influence of the Decentralized Finance (DeFi) movement. Now, the Blockchain community had a token-based fundraising method that was both exchange-based, and decentralized in nature. This made the funding process completely open to everyone who had a crypto wallet. Plus, the fees were a lot cheaper on DEX's because they weren't centralized profit driven businesses.

Then, the Decentralized Finance (DeFi) movement gave rise to the Game Finance (GameFi) movement. GameFi added the financial power of tokenomics to the robust world of gaming. This created a decentralized economic gaming model that enabled gamers to benefit financially from their gaming activities. In legacy gaming models the centralized gaming corporation that owns the game is the only one that is allowed to make money. The players do not benefit from the money, and time that they invested over the years. But, in GameFi the players earn by playing the games, and they get to own any game assets that they have acquired during playing the game. In GameFi the players earn tokens associated with each game. Both these tokens, and game assets in the form of NFT's can be converted to cash, or other digital assets on an ongoing basis. This gave rise to the next form of token-based fundraising. The GameFi movement created the Initial Game Offering (IGO's) method to fund the startup of new Blockchain-based games. Plus, you do not have to be a game player to benefit from this game-based token model. GameFi is another major aspect of the growing decentralized finance movement.

Currently, the most innovative form of token fundraising is the Initial Liquidity Offering (ILO's) method. One issue that affects both newly launched tokens, and DEX's is the lack of proper liquidity. Tokens need to be back by some form of liquidity to render it valuable. And, exchanges need an abundance of varied tokens to remain liquid, and valuable. What good is an exchange if they do not have the tokens you need, or enough of the tokens that you desire. Sometimes early investors will sell off newly launched tokens in their portfolio for the quick profit returns. When this happens the new token loses a lot of its liquidity. This also creates liquidity issues for the exchange as well due to the instability of the token in the market. An ILO provides solutions for both the token, and the exchange liquidity issues.

The concept of an ILO builds on the DeFi movement's progress by working with the decentralized exchanges. The ILO is a mechanism that funds projects by pairing the new project token with an already established liquidity tokens such as Ethereum, or other stable coins. This is how this works...the newly minted tokens are paired with an acceptable established easily tradable token of value. Both tokens are deposited into a liquidity pool on a decentralized exchange. These liquidity pools are open to all liquidity investors on the exchanges. The exchanges use these liquidity pools to facilitate token trading for the crypto community when they need certain tokens. These trades generate fees for both the DEX, and the liquidity pools. This means the ILO launch locks up a certain amount of its newly minted tokens with valued tokens in fee generating exchange pools, and receives an ongoing percentage of the profits according to the investment in the liquidity pool. This fund-raising method provides liquidity for the new project token, and an initial listing on exchanges. It also gives the project a safe mechanism for investors to build the value of the new project's token. Plus, this method is a way to immediately start generating income that can be used for the growth of the project.

Now that you understand the basic methods of token-based fund-raising methods, lets examine some of the ways to launch a new token for your project. An entire industry has spawned for token launching in the past few years. The community calls these conduits for creating a new token crypto-launchpads. There are numerous launchpads in the Blockchain ecosystems. I will list some launchpads for ILO's, IDO's, and IGO's. Here are some starting points to begin your research:

Initial Token Launchpads

- Uniswap
- Uncx.network
- DAO Maker
- Polkastarter
- PancakeSwap
- Coinlist
- Enjinstarter
- Honeypot Finance
- Impossible Finance
- Kommunitas
- Moonstarter
- WeWay
- GameFi
- BullPerks
- Solanium

Here are some good places to start your research: coinlaunch.space, cryptorank.io, and coinlist.co. A basic search for IDO token launchpads will definitely yield sufficient results. There are a lot of options so do your research, and due diligence before making your choice.

Another valuable resource that is available for funding business are business grants. Business grants are funding sources that are awarded to individuals, and companies that do not have to be repaid like a business loan. Grants to businesses are rewarded for specific business purposes. The grant awards can be for seed funding, operations, research, specific solutions, or by identity-based demographics among other business reasons. Business grants do not have the financial stipulations of a business loan. And, grants do not sacrifice business equity, or ownership in your business in return for funding like venture capital funding.

In the past decade the Blockchain grant funding industry has grown tremendously on a global scale. There are Blockchain specific grants, project-based grants, and solution-based grants. There are numerous grants for Blockchain developers in the crypto ecosystem. Take time to research the various types of Blockchain grants that are available to aspiring entrepreneurs. See what type of Blockchain grants align with your particular business goals. Then check the grant provider's criteria, and application processes. Be aware that most grants have a time-specific deadline for submitting applications for funding. This means a major part of your grant researching process will be forming timelines, and funding cycles for each grant opportunity. Also, keep in mind that the application, and selection process can be extremely competitive due to the fact that this funding method comes with fewer strings attached than a business loan, or venture capital deal. There are hundreds of Blockchain grant providers so a basic online search can jump start your deep dive into this funding avenue.

Seeking funding from a venture capitalist is a totally different process. And, it is a process that not many people can give you genuine insight, and guidance to produce a successful outcome. Do not be fooled by goofy TV shows like the Shark Tank. Anytime you are asking someone for money it is not a game. Especially when you are asking total strangers to bankroll your dream job, or a life changing idea that will take sacrifice to manifest. Seeking the needed funding from a venture capitalist is basically determining how much ownership you are willing to sacrifice for some business capital. Giving up equity means giving up ownership of the company that you are still seeking to build. So, if you are dividing ownership with partners, and staff then giving up additional equity means more slices into that pie. And, venture capitalists are not your friends, and they do not share your emotional attachments to the project. Providing funding, and getting huge returns is their business model. Those returns come in the form of partial business ownership, and profits. Seeking funding from a venture capitalists means deciding how much of your future capital are you willing to part ways with to get someone to part ways with their current capital. And, both sides will have several needs that must be met to complete this process.

Venture capitalists make money on the equity that they acquired, and they can sell that equity for a profit to someone else, or they can hold onto that equity forever. Investors tend to have different timelines than original business owner's timeline. So, before you pitch your idea to venture capitalist make sure you have your game on point. You do not want to end up being a virgin asking a pimp for financial assistance. The investor is not the person who has to do all of the daily hard work, but they are the ones who will benefit financially from the equity that you have given up in exchange for their capital. Study the process of applying for capital before you request funding from others. Make sure you understand how to calculate your valuation because this will determine how much equity that an investor feels they deserve in your business. And, research the actual venture capitalist as much as you would study the terms of the deals offered to you. Cultivate a strong startup mentality to prepare yourself, and your team for the future phases of development. Build out your pitch deck, and cultivate your communication skills. Make sure you can pitch your entire business plan to anyone at anytime in 3 minutes tops. Then make sure all of the paperwork is in order for everyone involved in your business venture. Please, take the required time needed to get your goals accomplished properly. Good Luck My Friend!!!

CHAPTER XII

Final Thoughts On Blockchain Banking

It's October 2024, a few weeks before the U.S. presidential election, and I'm putting the final touches on this book. The financial landscape is extremely precarious. I'm living in a disaster stricken area that was severely impacted by one of the worst hurricanes to hit the eastern coastal region. It's been about a week since the power was restored, and two weeks since we removed the fallen tree from our roof. I just received an email from the local financial institution that most of their locations, and ATM's are back in service for the customers. Local banks have been out of service for a half a month for its most vulnerable customers during life threatening situations. Family legacies are being destroyed, and legacy banking was incapable of providing their customers their own money to manage their survival needs. A few days ago Bank of America's customers couldn't login into their accounts. Some people were able to login into their accounts, but all of their accounts said $0 available. Other BOA customers were told they did not even have an account with the bank. BOA just told the nation it's just a glitch. This is how one of the biggest banks in the world responded to millions of people who lost almost everything. Blockchain technology never stopped transactions from being processed during these traumatic times in people's lives.

We are a few weeks from a national election, and both presidential candidates are financially inept administrators. One candidate is part of an administration that engineered the worst inflationary policy the U.S. has ever seen. That person has already signaled that nothing is going to change if they are elected because, "bidenomics is working". This administration is totally against Blockchain, but they have unwavering support for multiple foreign genocidal wars. They have no concern, nor acumen to empower their own citizens, but can't wait to give damn near a hundred billion dollars to dictators. Any politician who feels like they do not need to run their campaign on actual issues, policies, or public interviews will never represent you as a citizen. These politicians do not even feel they should be accountable to the public during their campaigns let alone when their in their office. This is the reason that certain politicians are against Blockchain because the technology would make them accountable in all of their financial dealings. This would hold true for both domestic, and foreign budget policies.

The other candidate has gone bankrupt so many times, and exhausted all of his domestic banking options to the point that they had to beg for a loan from a foreign bank. The foreign bank that loaned this presidential candidate billions was the same bank that loaned Hitler millions!!! It was the bank formerly known as the Reich Bank of Germany. We call it Duetsche Bank today. How can a nation, and it's citizens be the home of the free when their national leaders are indebted to foreign legacy banks. All global legacy banks have national intelligence agency relations that run extremely deep into their organizations. To bad politicians are not required to conduct their financial transactions on the Blockchain. Then, we would be able to see what these backdoor financial deals really mean for our nation. On top of a foreign bank's financing of an U.S. candidate, we that same candidate announcing that he is building a blockchain financial platform with his corrupt family. It is more proof that no matter who you are you can use Blockchain technology when your access to the banking system has been restricted.

Those are the two presidential candidates, and their financial stances in regards to Blockchain. But, what about the president that is currently in office...you know the dead zombie that everyone pretends is lucid. Well, one of his first acts in office was to push for a 40% tax on Blockchain in the form of domestic capital gains tax. This announcement came right after the government labeled crypto transactions as capital gains. But, at the very same time he was in the public eye encouraging one of the most racist nazified nations to fund their war through Bitcoin until he could re-distribute American tax payer money to their foreign coffers. So to be clear, this person restricted, and stole wealth that was generated domestically through Blockchain technology, and encouraged foreign governments to embrace the technology. He told Ukraine get you money up with Bitcoin, and told U.S. citizens you'll be punished for using Blockchain technology.

In addition to, legacy banks failing the people during disasters, and politicians failing the people all of the time we have a nation of CEO's resigning at record paces for participating in the child rape, and forced sodomy networks that are used to compromise the wealthy, powerful, and those that desire these affiliations. You have a candidates that have been documented in these satanic cult networks. The U.S. economy is at its weakest point, and the captains of industry are all compromised. Legacy banks are closing branches at alarming rates. And, politicians are incapable of leading the people out of crisis.

Not to mention BRICS is leading the charge against de-dollarization. Major global nations, economies, and regional organizations are conducting transactions in non-U.S. currencies. Saudi Arabia refused to renew the petro-dollar agreement. Even Nigeria announced that it would start conducting oil transactions in their own currency. Russia, and China conduct all of their trade in their own currencies. Their are a handful of nations that have abandoned paper fiat altogether. So, you can be as skeptical as you want about Blockchain technology, but

I assure you no economy will emerge from crisis without embracing Blockchain. You can wait on your trusted leaders to guide you towards this path, or you can become one of the new leaders of this new economy called Blockchain.

APPENDIX PART 1- RESOURCES & LINKS

Useful Resources To Check Out

1. Fiatleak Map- https://fiatleak.com/
2. M1 Money Supply- https://fred.stlouisfed.org/series/M1SL
3. Coingecko- https://www.coingecko.com/
4. Coin ATM radar- https://coinatmradar.com/
5. CoinMarketCap- https://coinmarketcap.com/
6. DeFiLlama- https://defillama.com/
7. DeFi Pulse- https://www.defipulse.com/
8. Rarible- https://rarible.com/
9. Cointelegraph- https://cointelegraph.com/
10. Dune- https://dune.com/home
11. Dapp.Expert- https://dapp.expert/
12. RugDoc- https://rugdoc.io/
13. DappRadar- https://dabbradar.us/
14. Alphaday- https://www.alphaday.com/
15. Alchemy- https://www.alchemy.com/
16. Debank- https://debank.com/
17. Nansen- https://www.nansen.ai/
18. Etherscan- https://etherscan.io/
19. Coindar- https://coindar.org/
20. Messari- https://messari.io/
21. Coinigy- https://www.coinigy.com/
22. CryptoPanic- https://cryptopanic.com/
23. Coin360- https://coin360.com/
24. Coinidol- https://coinidol.com/
25. DeFi Rate- https://defirate.com/

APPENDIX PART 2- THE DEATH OF BANKERS LIST

Deaths Of Legacy Bankers Listing

In the past decade there has been over 100 banker deaths that have occurred in relation to scandals, investigations, and cover-ups. Some of these banker deaths are related to the global fraud practices of the legacy banking industry. Other banker deaths are related to Donald Drumpf's finances at Deutsche Bank(formerly known as the Reich Bank). Some bankers were killed without the public knowing the true reason of their untimely deaths. Here is a brief list of 80 bankers who have died in the past 10 - 12 years:

1. Stuart Smith- Senior Vice President of Operations, New York Mercantile Exchange= Disappeared
2. Fang Fang- JP Morgan China= Disappeared
3. Thomas James Schenkman- Managing Director of Global Infrastructure, JP Morgan= Cause of Death Unknown
4. Ryan Henry Crane- Executive, JP Morgan= Sudden Death
5. Richard Gravino- Application Team Lead, JP Morgan= Sudden Death
6. Jimmy Lee- Vice Chairman, JP Morgan Chase= Death Unexpected
7. Annie Korkki- JP Morgan Chase= Found Dead With Sister in Hotel Room (excess fluid in lungs & Cerebral Edema)
8. Robin Korkki- Head of Forex & Metals at Allston Trading= Found Dead With Sister in Hotel Room (excess fluid in lungs & Cerebral Edema)
9. Jason Alan Salais- Information Technology Specialist, JP Morgan= Found Dead Outside of Walgreens
10. Graeme Porteous- Investment Banker, JP Morgan & UBS Investment Banking Mining & Energy= Skiing Accident
11. Joseph A. Giampapa- Corporate Bankruptcy Lawyer, JP Morgan= Run Over by a Van

12. Ezdehar Husainat- Banker, JP Morgan= Run Over by a SUV
13. Aditya Tomar- Vice President of Technology, JP Morgan, Sanford C. Bernstein & CO., Barclays Capital= Died in N.Y. Train Crash
14. Kenneth Bellando- Investment Banker, JP Morgan= "Suicide" by Jumping Off His Apartment Building
15. Gabriel Magee- Vice President of Technology, JP Morgan= "Suicide by Jumping Off JP Morgan's Headquarters in Europe
16. Li Junjie- Financial Trader, JP Morgan= "Suicide by Jumping Off JP Morgan's Headquarters in Hong Kong
17. Joseph Nadol- Financial Aerospace Analyst, JP Morgan= Run Over by a SUV
18. Julian Knott- Executive Director of Global Tier 3 Network Operations, JP Morgan London= "Shot Wife with Shotgun Multiple Times Before Shooting Himself"
19. Joseph Ambrosio- Financial Analyst, JP Morgan= Died Suddenly from Acute Respiratory Syndrome
20. Thomas Bowers- Head of US Wealth-Management, Deutsche Bank= Suicide by hanging (**Managed Trump portfolio**)
21. William Broeksmit- Risk overseer, Deutsche Bank= Found hanging by his dog's leash
22. Susan Hewitt – Vice President, Deutsche Bank, Drowned in a stream
23. Calogero Gambino- Associate General Counsel and Managing Director at Deutsche Bank, America= Alleged SUICIDE by hanging
24. Jeffery Epstein- Limited Partner at Bear Sterns and Liquid Funding LTD= "Suicide at Metropolitan Correctional Center"
25. Venera Minakhmetova- Former Financial Analyst at Bank of America Merrill Lynch= Hit by a truck while cyclin
26. Michael Burdin- Bank of America , Foreign Exchange Manager= Suicide by "jumping in front of train"
27. Chris Latham- Executive, Bank of America= Convicted of Murder for Hire because hitman confessed during a traffic stop
28. Keith Barnish- Former Senior Vice President, Bank of America & Senior Managing Director at Doral Financial Corporation= "Died Suddenly"
29. Bruce Steinberg- Former Hedge Fund Executive, Credit Suisse & Bridgewater= Died in Plane Crash with Family
30. Dan Hanegby- Investment banker from Credit Suisse Group= Crushed by NYC Bus
31. Tanji Dewberry- Assistant Vice President, Credit Suisse= Died in House Fire
32. Jennifer Riordan- VP of Community Relations, Wells Fargo Bank= Passenger that was sucked halfway out of the plane's broken window mid-flight after a piece of the engine struck the window where she was sitting on a Southwest Airlines flight from New York to Dallas

33. Abid Gilani- Senior VP, Wells Fargo Bank= Died when the brakes on a train failed forcing the train to enter a turn at 100+ miles per hour
34. Patrick Sheehan- Managing Director of Finance, Wells Fargo Bank= Died in car accident
35. Nicholas Valtz- Managing Director, Goldman Sachs= He was found floating with his kite board
36. Vusi Mhlanzi- CEO, Basis Points Capital= Shot 6 times at a busy traffic intersection
37. Martin Senn- Former CEO, Zurich Insurance= "Shot himself in his Swiss resort home"
38. James McDonald- President & CEO, Rockefeller & Co= "Shot himself"
39. Miguel Blesa- Former Chairman, Caja Madrid= Died from being shot in the chest at a private hunting lodge
40. Aleksandr Potyomkin- Director, Russian Central Commercial Bank= Shot in his apartment stairwell
41. Thomas Gilbert Sr.- Founder, Wainscott Capital Partners= Shot inside of his bedroom
42. Hussain Najadi- CEO & Chairman, AIAK Group= Executed in front of his wife outside of his banking group's headquarters
43. Jan Peter Schmittmann- Former CEO, Dutch Bank ABN Amro= Murder-Suicice involving wife and daughter
44. Benedict Philippens- Director & Manager, Bank Ans-Saint-Nicolas= Executed with his wife and a child-related relative
45. Juergen Frick- CEO, Bank Frick & Co. AG= Executed in his banking firm's garage by the man he sued for extortion 3 years prior
46. Mohamed Hamwi- Financial System Analyst, Trepp= Shot dead at a busy New York intersection...(He worked at the World Trade Center, but was late to work on 9/11)
47. Naseem Mubeen- Assistant Vice President, ZBTL Bank= Suicide jumped off the tenth floor of the bank
48. Mike Dueker- Chief Economist, Russell Investments= "Suicide by jumping over a 4 foot fence so he can jump another 50 - 60 feet to his death
49. Lydia- Banker, Banque-Populaire= "Suicide by jumping from 14 story window"
50. Edmund Reilly- Trader, Vertical Group= "Suicide by throwing himself in front of a train"
51. David William Waygood- Former Banker, HSBC= Suicide by stepping in front of a train
52. Thomas J Hughes- Investment Banker, Goldman sachs= Suicide by jumping from his 14th floor luxury building into moving traffic
53. Kenneth Ballando- Investment Banker, Levy Capital= Suicide when he jumped from the 6th story of his apartment building

54. Robert Wilson- Retired Hedge Fund Founder, Wilson & Associates= "Suicide when the 87 year old jumped to his death from his 16th floor residence

55. Pierre Wauthier- Chief Financial Officer, Zurich Insurance Group= Suicide at home

56. David Rossi- Communications Director, Monte dei Paschi di Siena (MPS) Bank, Thrown off top of bank building, video shows him facing building upon impact, and then suspected killers check his body in the alley

57. Thierry Leyne- Banker, Anatevka S.A., Israël= Suicide, jumped from the 26th floor

58. Nuno Ribeiro da Cunha- Private Banker, EuroBic= Suicide, 2nd attempt was successful (He was personal banker to richest woman in Africa)

59. Aivar Rehe- Former CEO, Danske Bank= Found dead during bank investigation

60. Adrian Hill- Former Chief Financial Officer, HFC Bank= "Drowned himself in his private pool"

61. James Erven- Head of Digital Development & Retail and Business Banking, Santander PLC= Suicide by jumping off the balcony's of the top floor of a building

62. Michael Treichl- Co-Founder, Audley Capital Advisors= Found dead in a lake near his residence

63. Neil Rodney Smith- Founder & CEO, Infraccess= Found in an apartment with a wet towel stuck in his mouth, and a pillow over his head

64. James Starkey- Banker, Catalyst= Found tortured, and murdered in South African apartment

65. Oliver Dearlove- Relationship Manager, Duncan Lawrie bank= Died from being physcially attacked on the street

66. Alex Lagowitz- Futures trader with Bank of America/Meryll Lynch= Found dead after falling from the 23rd floor of his residential building

67. Roger Agnelli- Former Banker, Vale= Died when his plane crashed into residential homes shortly after takeoff

68. Shawn Miller- Managing Director, Citigroup= Found murdered in bathtub with throat slashed

69. Omar Meza- Board Executive, AIG Financial Distributors= Found dead in a Marriott Hotel pond after being missing for a week

70. Nigel Sharvin- Senior Relationship Manager, Ulster Bank= Found drowned after attending a stag party

71. Daniel Leaf- Senior Manager, Bank of Scotland/Saracen Fund Managers, Fell to his death over 1000 feet from a cliff

72. Tim Dickenson- Communications Director, Swiss Re AG= Found dead/cause unknown

73. Geert Tack- Private Banker, ING= Body found off the coast of Ostend, the body was remove from water before discovery

74. Therese Brouwer- Managing Director, ING=Died in MH17 Crash

75. Tod Robert Edward- Vice President, M&T Bank= Killed when hit by a truck
76. Amir Kess- Co-Founder & Managing Director, Markstone Capital Group= Killed when hit by a car while biking
77. Benjamin Idim- Banker, Diamond Bank= Died in car accident
78. Andrew Jarzyk- Assistant Vice President, Commercial Banking at PNC Financial Services Group= Body pulled from the river
79. James Stuart Jr.- Former CEO, National Bank of Commerce= Found dead
80. John Ruiz- Municipal Debt Analyst, Morgan Stanley= Died suddenly

This list of banker deaths keeps growing well into 2024 like a contagious fungus. Trying to list all of their names would require its own book. I believe a true deep dive into the consistent removal of bankers will provide much needed clarity into the coming collapse of the legacy banking system. A follow up is definitely warranted, in order to, expose the truth about how our society truly operates in times of crisis. Good Luck to all of the bankers, and future bankers out there!!!

I have been studying, and researching Blockchain for almost 15 years. I have a few Blockchain Certifications, and a consulting company based in Estonia. I have been a member of a few Blockchain organizations. Now, I'm focused on building a web3 Blockchain portal to onboard citizens.

RABBIT HOLE CONSULTING

BANKING THE UN-BANKED